Tyler Grosskopf

BASIC
DRAFTING
a manual for beginning drafters

Written and illustrated by Leland Scott
Cover created by Chris Scott

Note for Librarians: A cataloguing record for this book is available from
Library and Archives Canada at www.collectionscanada.ca/amicus/index-e.html
ISBN 1-4120-9676-6

Printed in Victoria, BC, Canada. Printed on paper with minimum 30% recycled fibre.
Trafford's print shop runs on "green energy" from solar, wind and other environmentally-friendly power sources.

TRAFFORD
PUBLISHING™

Offices in Canada, USA, Ireland and UK

Book sales for North America and international:
Trafford Publishing, 6E–2333 Government St.,
Victoria, BC V8T 4P4 CANADA
phone 250 383 6864 (toll-free 1 888 232 4444)
fax 250 383 6804; email to orders@trafford.com
Book sales in Europe:
Trafford Publishing (UK) Limited, 9 Park End Street, 2nd Floor
Oxford, UK OX1 1HH UNITED KINGDOM
phone 44 (0)1865 722 113 (local rate 0845 230 9601)
facsimile 44 (0)1865 722 868; info.uk@trafford.com
Order online at:
trafford.com/06-1432

10 9 8 7 6 5 4 3 2

About the author:

Leland Scott has been an adjunct faculty member at Baker College of Owosso for twelve years, teaching both board and CAD drafting. In 2002, he was named Instructor of the year. In addition, he owns and operates his own design business.

Leland received an MBA at Baker College, a Bachelor's degree from John Wesley College, a vocational education teaching certificate from the University of Maryland, Teaching certificate from the U.S. Navy, and diplomas from two Navy drafting schools. Leland also attended William and Mary College, Cecil Community College, Old Dominion College, and Lansing Community College. Leland served as a Draftsman and an instructor in the U. S. Navy for 20 years; worked, and taught, in three engineering companies as well as teaching in Lansing Community college.

Leland's experience has been in all phases of drafting and field work, including machine drafting, architectural drafting, civil drafting and graphic arts, covering nearly fifty years experience as a draftsman, technician, inspector, and instructor.

CONTENTS

Basic Drafting   1

Preface

Introduction   3

History   5

Purpose   7

Drafting Instruments   9

Drafting instruments
    The drafting board   9
    The T Square   10
    Straightedge or Parallel Ruler   11
    Triangles   12
    Drafting Pencils   13
    Lead Sharpeners   15
    Pens   16
    Drafting Paper   16
    Drafting Tape   16
    Erasers   17
    Erasing shield   17
    Lettering guide   18
    Compass   20
    Dividers   22

Curves 23
Protractor 25
Scales 26
Care of instruments 27

Techniques 29

Setting up the board 29
Instrument layout 30
Lettering 31
Guidelines 32
Title Block 33
Drawing lines 36
Geometric Techniques 37

Orthographic Projection 43

Views 43
Relationship of views 48
Primary views 48
Practice Exercise 50

Getting started 51

Planning the drawing 51
Laying out the drawing 51
Drawing procedure 53

Drafting Conventions 57

Alphabet of lines 57
Symbols 59

Measuring 61

Engineering scale	61
Metric scale	63
Architect's scale	63
Measuring practice	65

Dimensioning 67

Extension lines	68
Dimension lines	68
Arrow heads	68
Dimension	68
Unidirectional	69
Aligned	69
Fractional	69
Decimals	70
Arcs and Circles	71
Architectural	71

Sectioning 73

Cutting plane	73
Cutting Plane Line	74
Arrow head	74
Direction of view	74
Types of sections	
Full section	75

Half section   75
Offset section   76
Aligned section   76
Removed section   77
Revolved section   77
Broken out section   78
Section of mating parts   79
Sectioning Exercises   80

Pictorial Drawings   **81**

Perspective drawings
 One point perspectives   82
 Two point perspectives   83
Oblique drawings   85
 Cavalier drawings   86
 Cabinet drawings   87
Isometric drawings   88

Sketching   **93**

Drawing Exercises   **101**

Appendix   **135**

BASIC DRAFTING

Preface:

This book is intended to help the beginning drafter understand the basics of the drafting trade. It will provide a good base for any type of drafting career. Beyond this beginning, lies many choices to build on, but the basics are always the same. It is expected that the drafter will, at some time, advance to computer aided drafting, but the skills learned on the board will remain as a foundation of understanding the principles of all that is drafting whether on the board or computer. These principles have stood through time and are the basis of all drafting communication.

The book could easily be used either as a classroom text for beginning drafters, or by someone who would take the time to learn the trade on their own. It progresses through the basic terms and definitions of instruments to instructions for drafting techniques and provides some practice assignments. The appendix provides some further information that would be useful to any drafter.

The book is noticeably without the usual "textbook format" - not the usual divisions of Chapters and arduous study, it is intended to be a casual and natural walk through the basics in a step by step process from the very basic information about materials and instruments to a procedural process of beginning the task of drawing various types of drawings. Beginning with an introduction to instruments and advancing to techniques for their use, prepares the drafter for the areas of orthographic projection, sectioning, dimensioning, and pictorial drawing. Sketching is left intentionally until the end of the book so that it may take advantage of the 'process of drawing' before attempting to sketch.

INTRODUCTION

INTRODUCTION

Drafting is used in every facet of business today. Whether it be in Machine drafting, Electrical Drafting, Aeronautical Drafting, Civil Drafting, Architectural Drafting, or the design of any other products, or graphic arts, it begins with one common beginning – visualize and convey. Everything that is manufactured, from the house you live in to the car you drive, the road you drive it on, and the pen you write with, started with an idea, was transferred to paper, and conveyed to the manufacturer. This means that drafting is a time-tested career that has an ever expanding future. The art of drafting has gone through several stages, but always remained the same in that it is an art of putting ideas on paper in a way that someone else can use these ideas to produce an end product. As we progressed through the beginnings of manufacturing, into the industrial age, and into today's more technological time, one thing has remained constant – this need to communicate ideas.

HISTORY

Thousands of years ago, men drew their ideas on parchment or papyrus sheets as they designed pyramids, temples, and structures of the day. Many of those structures stand today as testimony to the precision of planning that was transferred from thinkers to craftsmen. Moving into the age of invention and industrialization, more and more the need to convey new ideas was made evident. As civilization progressed, the drawing process became more organized and drawings were produced with more expertise and more as an art by professionals who were trained and hired for the purpose. These professionals came to be known as 'draftsmen'.

The equipment used to produce hand drawings has changed little over the years. The basic drafting instruments have become more refined to give more accuracy and make drawing easier, but remain basically the same. A drafter requires a good surface on which to draw, some basic instruments to aid in giving smooth, accurate lines and a medium on which to draw.

Many types of mediums were used, all trying to provide a long lasting document that could be stored, retrieved, and read over and over again - parchment, papers, and fabrics. It was discovered that Linen could be starched and pressed to provide a writing/drawing surface suitable for pencil, and especially ink. Although expensive, this linen would withstand the repeated abuse of rolling, folding, and storage. It was difficult to make changes on linen because whenever the starched surface was roughened by an eraser, it had to be resurfaced before it could be used again. Special surfacing compound was available for that purpose, but it was time-consuming and less than perfect. Various papers were developed to replace the linen. Thin papers would allow tracing but had no longevity. Many papers became brittle and discolored with age. All materials had to have an ability to accept ink or pencil and the durability to withstand time. As plastics entered the scene, a plastic film was developed commonly called 'Mylar' – a trade name. Mylar was virtually indestructible. Although it could not be folded, it could be rolled or stored flat. It required a 'plastic' lead for good pencil image that was not easily smudged. It inked well and was easy to change or correct without damage to the material. For reproduction with the diazo blueline process, the material must be translucent – have a quality that allows light to pass through it. Mylar was very translucent and could easily be used for tracing and for reproduction with the Diazo process. Mylar, also, could be 'layered' – a process of drawing different 'disciplines' or features on separate sheets and then mating the sheets on top of one another – sometimes as many as four or five sheets could be run together through the reproduction process with very little distortion or image loss. Equally popular, because of its availability and cost, is Vellum - a high grade of tracing paper. With the arrival of the computer and copiers, a heavier paper is used for direct process prints, but hand drawn originals are still done on Mylar or Vellum.

Computers and Computer Aided Design are the new medium of drafting. It is an exciting change to many of the old methods, but in the end, a printed product is still required. In the process of Computer Drafting, hand drawing lies in the foundation of understanding. Computer drawings are often preceded by sketched drawings. Computer Aided Drafting allows the drawings to be taken to new dimensions of 3-D and animation, however, working drawings still must be produced to give the information to the manufacturer.

PURPOSE

PURPOSE

Technically speaking, the drawing has only one purpose – to convey to the manufacturer, the dimensions, and instructions for production. More ideally, it conveys the feeling of the designer, the accuracy, and the appearance of the end result. In a machine shop, the machinist must understand the use, and sometimes, urgency of the end product to give proper attention to detail and accuracy. A part for a space shuttle would require much more precision than the handle for a garden tool. As the machinist understands this precision, he/she will apply the proper attention to detail. This information is, of course, a part of the job specifications, but is conveyed in the drawing prepared in the drafting room. As the engineer, architect, or designer, relates this information to the drafter, great care must be given to insure that every detail is portrayed correctly.

There are many types of drawings. As mentioned earlier, some stand alone as plans for the manufacture of an object - or a part of some larger project. Drawings may be a part of a set of drawings for the assembly of a large piece of equipment, an auto, aircraft, or for something like a toy or piece of jewelry. They may be for a building to be constructed, a bridge, or roadways to be built. To properly appreciate the need for the drawing it is important that the drafter envision the end result. A good drafter develops this ability early in his/her career and hones the skills to match. A good drafter remembers that every drawing is a reflection of the drafter and his/her professional image.

DRAFTING INSTRUMENTS

DRAFTING INSTRUMENTS

The instruments used by a drafter depend upon the type of drawing that is going to be made and to what extent of accuracy those drawings are to be drawn. Most drawings require very unsophisticated instruments. Listed below are some of these basic instruments with their description and use.

The Drafting Board

The drafting board is a portable 'table' and can be used both in the office, and in the field. The board can be any size - a rectangle large enough to accommodate the size drawing paper to be used. It can be made of any material, but is usually manufactured of a cored wood with a veneer or plastic surface. The board must provide a smooth surface free of raised or indented imperfections that would hinder the drafter in drawing smooth, even lines. It should have a smooth edge for the head of a T Square to ride along, or have an attached parallel ruler to create parallel horizontal lines. The board is often 'covered' with a plastic surface that provides a good drawing surface and is easily cleaned. This cover is a temporary surface that is attached to the board with double faced tape and is easily removed or replaced.

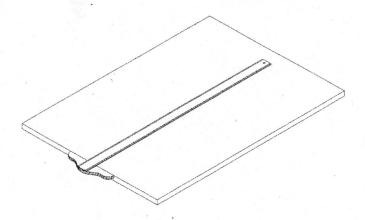

The T Square

The T Square is a tool used to draw horizontal lines and to be the base for other instruments. The T Square has two parts – the head, and the blade. The two are joined at right angles to form a T. The head rides along the edge of the board and the blade extends across the board. The head must be kept firmly against the board as the T Square is moved up and down the board to position it for drawing lines. Keeping it snugly against the board assures that the blade remains horizontal and true. The pencil is pulled against the top edge of the blade to draw horizontal lines. The T Square should be given utmost care to keep it from damage. The edge of the T Square blade should be kept free of nicks or dents to assure that it will always provide a straight edge for a good line. The head of the T Square should be protected to prevent it being moved out of 90° alignment with the blade. To avoid accidentally bumping the head out of alignment, it is best to store a T Square by hanging it by the hole in the end of the blade.

T Squares come in different lengths to accommodate the board chosen.

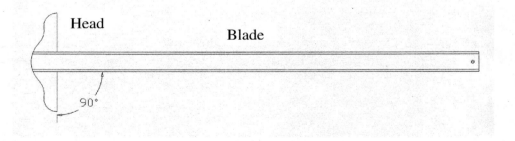

Head

Blade

90°

Straightedge or Parallel Ruler

The straightedge replaces the T Square. It is sometimes called the Jacob's Straightedge or Parallel Ruler. It is permanently attached to the board or table top and moves up and down the surface on a system of cords that keep it always true parallel. It is used to draw horizontal lines and to support the triangles. Like the T Square, the parallel ruler should be kept free from damage to assure a smooth even edge for accurate lines.

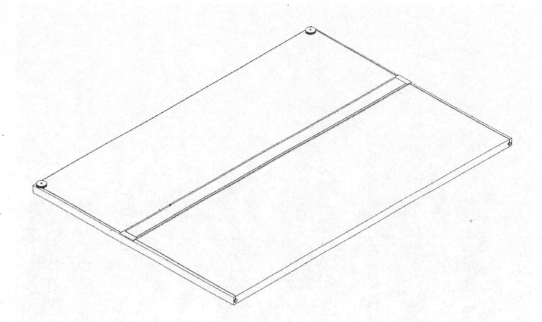

Triangles

Triangles are clear plastic tools used to draw straight precision lines at other than horizontal direction. They are accurately made to form angles of 90°, 45°, 30°, and 60°. This is done with two triangles. One is called, because of its angles, a '45'. It has two 45° angles and one 90°. The other is called a '30/60' because it has the angles of 30°, 60°, and 90°. The combination of these two triangles can form many other angles. Triangles come in many different sizes for ease of use in small or large drawing situations. The triangle is placed against the straightedge or T Square and the pencil is drawn along it to draw lines at the desired angles. Triangles should be kept free of nicks along their drawing edges to assure smooth accurate lines.

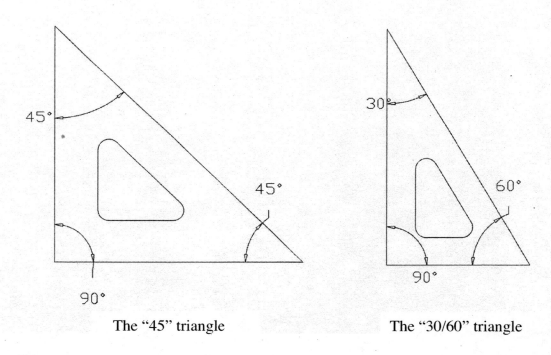

The "45" triangle   The "30/60" triangle

Drafting Pencils

Drafting pencils are selected for the desired result. Soft leaded pencils produce a dark line, but the line is easily smudged. Hard leaded pencil lines are less likely to smudge, but tend to produce a lighter line. It is important to have a good black line that can be controlled in width and stay sharp in appearance.

Leads range in hardness from 7B (very soft) to 9H (very hard)

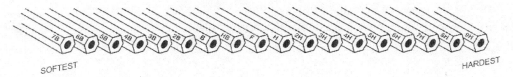

SOFTEST   HARDEST

Pencil Lead Hardness

Generally, a 4H pencil will give a good line for layout and construction lines, while a 2H will be a good pencil for final lines. Each drafter will develop a preference for the right pencil. Some prefer as hard as 7H for layout and H for finish lines. Some will use a softer leads such as HB for lettering. This will take time and experience to develop. There are many choices to select from as described here and illustrated on the next page.

It is important to keep pencils sharp to draw consistent lines. The finished line must be shiny black whether it is a thin line or a thick line. This is done by developing a consistent pressure on the pencil when drawing lines and lettering. A consistent line width is obtained by rotating the pencil as it is drawn along the straightedge.

Drafting pencils come in different types – **Wooden Pencils, Lead Holders**, and **Mechanical Pencils**.

Wooden pencils are made of wood with different leads. A variety of pencils with different lead softness should be kept handy and sharpened.

Lead holders are pencil shaped holders for standard pencil lead. A lead holder may be used for any softness of lead and the lead may be advanced or withdrawn to any length of lead exposed to be sharpened. The drafter usually keeps several different ones with different softness of lead handy and sharpened.

Mechanical drafting pencils use leads of various sizes. Each pencil is built to hold only one size lead. Lead can be of diameters of 0.3mm – 0.9mm and all of the variety of hardness described above. The mechanical pencil gives a uniform line size without sharpening. A drafter keeps several different pencils handy to be ready to change from one line to another without changing leads.

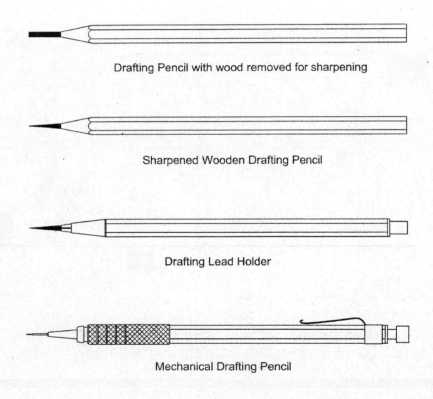

Drafting Pencil with wood removed for sharpening

Sharpened Wooden Drafting Pencil

Drafting Lead Holder

Mechanical Drafting Pencil

This illustration is an example of the different types of pencils described above.

Lead Sharpeners

Drafting pencils are not treated the same as regular writing pencils. The lead is sharpened to the desired point with a lead sharpener. The most common instrument for this is a Sanding pad - a pad of sandpaper on a wooden paddle. For wooden pencils, some of the wood is cut away to expose the lead for pointing on a sanding pad. Some mechanical pencil sharpeners are made for drafting pencils and remove only the wood, exposing the lead; or the wood can be carefully cut away with a sharp knife. This allows the lead to be sharpened on the sanding pad or lead pointer. When using lead holders, the lead can be extended the desired distance to be used against the sanding pad.

Another type of pointer is a mechanical pointer device. The pencil is inserted into the opening on the top, with lead extended, and as the device is turned, the lead is rotated against a sanding cone inside and sharpened.

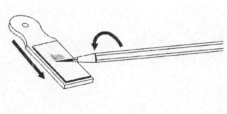

As shown at left, the pencil is rotated as it is drawn across the sandpaper pad to create a conical point.

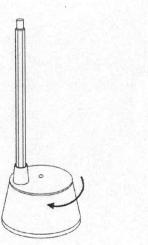

The mechanical pointer at left creates a conical point on the lead. The pencil is inserted as shown. The top of the device turns with the pencil, rubbing the lead against a sandpaper cone inside.

The sanding pad or any sharpeners should be kept away from the drawing to prevent stray graphite dust from falling on the drawing.

Pens

Another instrument for drawing is the Technical pen. Some drafters use Technical pens for inked work. Technical pens vary in width of point to give different line widths. This method gives a sharp black line that is basically permanent, without the hazard of graphite smudges. Using technical pens requires practice and concentration. It is important to keep the pen tip in a position that will not allow the ink to run under triangles or straightedges. Often the triangle or straightedge has an overhanging edge that allows the pen tip to ride vertically along it without the wet ink touching the edge and bleeding under it. Inked lines reproduce well, but require close attention to assure that every intersection meets precisely, and is correctly drawn the first time as they are hard to erase or edit.

Drafting Paper

As discussed earlier in the book, drafting is done on a variety of materials – usually vellum or Mylar. Vellum, a high grade of tracing paper, provides a good drawing surface, is relatively inexpensive and reproduces well. It is the favorite for most drafters today. Paper can be purchased in various standard sheet sizes and rolls. See the appendix for standard sheet sizes.

Drafting Tape

Drafting tape is used to hold the paper on the board. Drafting tape comes in rolls and in 'dots'. The dots are small round pieces of tape on a backing strip and are stored in a dispenser. They may be easily taken from the dispenser to hold the corners of the paper sheet to the board.

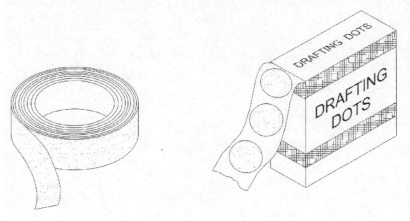

Erasers

Erasers are used to remove unwanted lines. Erasers come in several sizes and types. Art Gum erasers are a tan, soft rubber which will clean an area of smudges and light lines, but often do not have the power to remove stubborn marks and dark lines. The Pink Pearl has long been a favorite. It is a smooth, solid pink rubber eraser, much like the eraser on the end of a writing pencil, but may leave a pink color on the paper. White soft Plastic erasers are the most used today. They come in block shape and pencil shape. There are 'pencil' shaped erasers that have a paper body, with peel-off strips to expose the eraser, and there are mechanical erasers – in which refill erasers can be used, much like a lead holder.

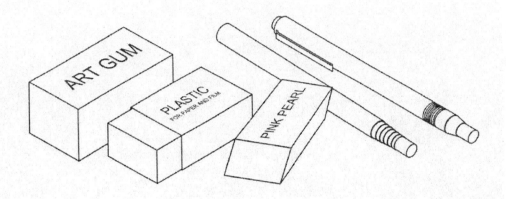

Erasing Shield

An erasing shield is just what the name implies – a shield to protect the drawing when erasing. The erasing shield is a thin metal plate with holes of various size and shape. The plate is laid over a feature that is to be erased and the erasure is made through the hole. The shield protects the rest of the drawing from the eraser.

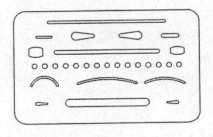

A typical Erasing Shield

Lettering Guide

Guide lines can be measured and drawn, but the task can be simplified with a lettering guide. The most common one is the **Ames Lettering Guide**. As pictured below, it is a simple looking plastic instrument, carefully engineered to assist in drawing properly spaced lines for lettering. It is adjustable for many lettering sizes. The guide is used by rotating the center disc to the desired spacing and moving the guide along the straightedge with a pencil point placed in the appropriate hole, drawing lines as it moves.

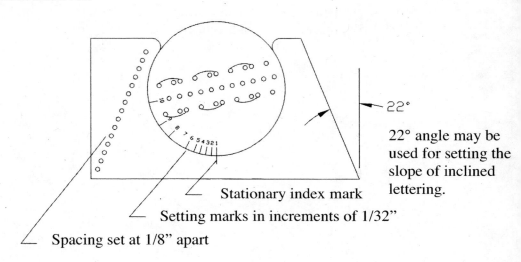

22°

22° angle may be used for setting the slope of inclined lettering.

Stationary index mark

Setting marks in increments of 1/32"

Spacing set at 1/8" apart

The Ames Lettering Guide

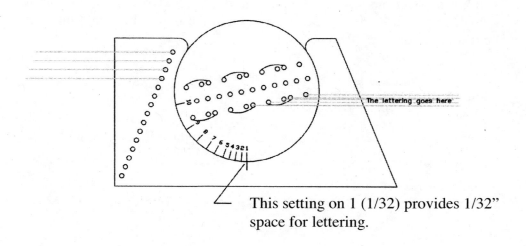

The lettering goes here

This setting on 1 (1/32) provides 1/32" space for lettering.

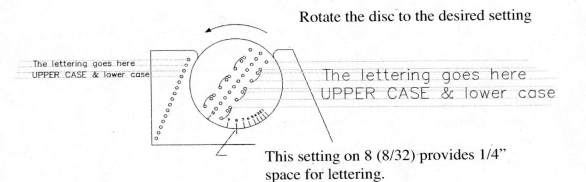

Rotate the disc to the desired setting

The lettering goes here
UPPER CASE & lower case

The lettering goes here
UPPER CASE & lower case

This setting on 8 (8/32) provides 1/4"
space for lettering.

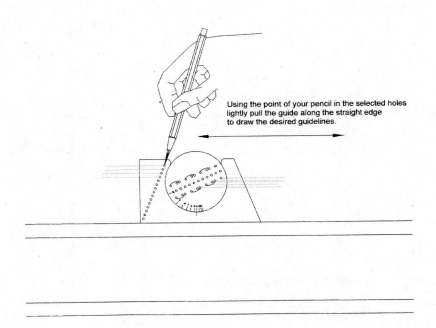

Using the point of your pencil in the selected holes
lightly pull the guide along the straight edge
to draw the desired guidelines.

Compass

The compass is an instrument used to draw circles. The compass has two legs – one with a point, the other with a lead. The compass is set to the desired radius by the use of an adjusting screw; then the point is set in place at the center of a desired circle and the lead is used to draw the circle by pivoting the compass around that point. The compass is a delicate instrument and should be carefully cared for.

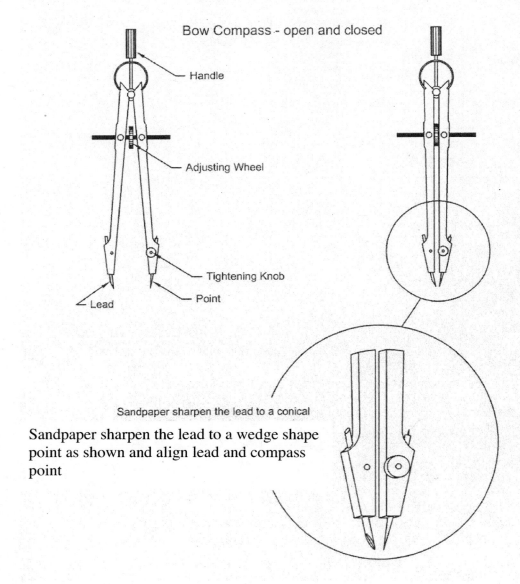

Bow Compass - open and closed

Handle

Adjusting Wheel

Tightening Knob

Point

Lead

Sandpaper sharpen the lead to a conical

Sandpaper sharpen the lead to a wedge shape point as shown and align lead and compass point

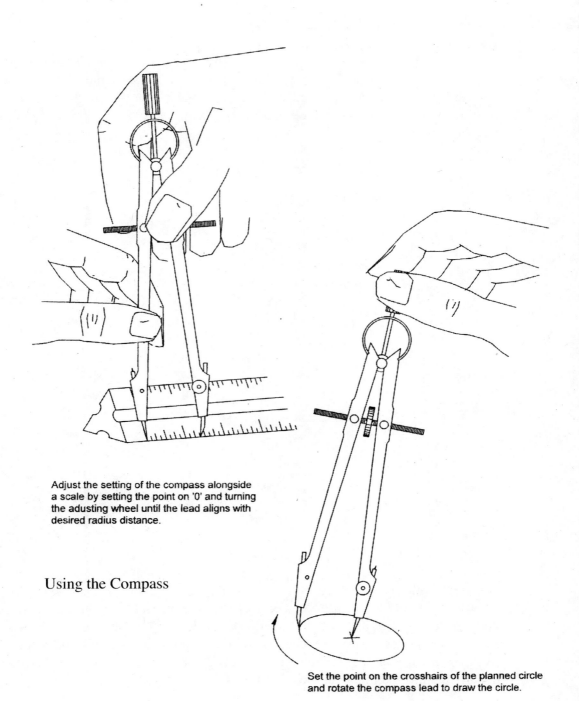

Adjust the setting of the compass alongside a scale by setting the point on '0' and turning the adusting wheel until the lead aligns with desired radius distance.

Using the Compass

Set the point on the crosshairs of the planned circle and rotate the compass lead to draw the circle.

Dividers

Dividers are similar to a compass in appearance, but have no lead, rather a point on each leg and are used to set off or duplicate distances. The legs are set to the desired distance and the points are used to give accurate points. The instrument should be well cared for to protect the points from damage.

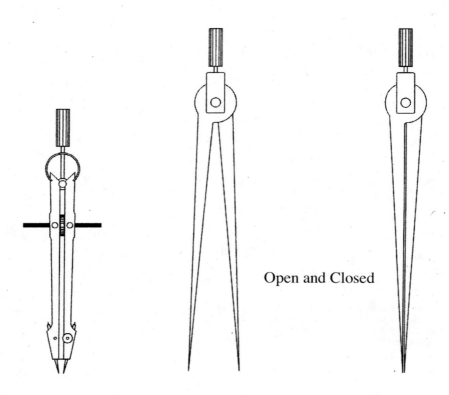

Open and Closed

Bow and Friction dividers.
The Bow divider is like the bow compass, except that it has two points and no lead. It can be adjusted with the adjusting screw. The friction dividers – sometimes larger – are adjusted by simply pulling the legs apart or pushing them together. The friction head holds them in place. They usually have fixed points at the end of each leg.

Curves

Irregular curves that cannot be drawn with a compass can be either drawn freehand or with the use of an instrument that gives a smooth, continuous line. This is done with 'Irregular' or 'French' Curves. French curves are plastic tools made up of many arcs of different sizes and lengths. French curves come in many shapes and sizes to accommodate almost any desired curve. The edges of the French curves should be protected from nicks to assure smooth even lines.

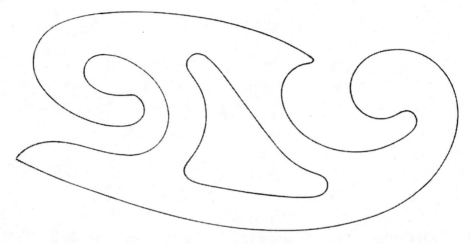

A typical French Curve

To use a French curve, the intended line needs to be laid out first. The line of the curve is laid out by either sketching the curve or mechanically locating points along the path of the intended curve. This can be done by calculating coordinates for certain points to give the general direction of the line. These points are then connected by selected sections of the French Curve. The line of dots is laid out first…

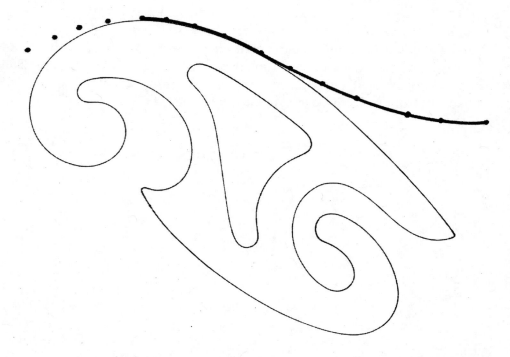

…Then the tool is placed along the dots until a section of the instrument aligns with at least three points. A line is drawn connecting these three points, then the instrument is moved to find another section that will connect another two or three dots, taking care to overlap the last two points of the line drawn. With some practice and much moving of the French curve, the resulting line will look like one continuous curved line.

Protractor

   The protractor is a half circle or full circle instrument divided into degrees. It is used to measure angles. Two sets of numbers on the protractor allow you to read either from right to left or left to right depending on the angle you are measuring.

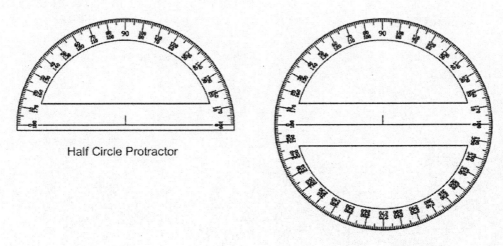

Half Circle Protractor

Full Circle Protractor

To use the protractor, the index mark of the protractor is set at the vertex of the angle with 0° set on one leg of the angle. The measurement is read along the arc of the protractor where it crosses the other leg of the angle.

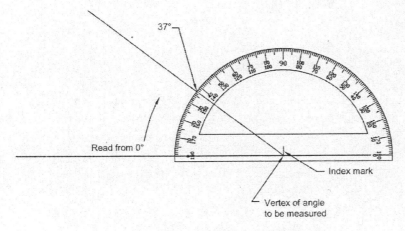

MEASURING AN ANGLE

Scales

Scales are measuring devices similar to rulers. Scales are set at different ratios to actual size. They get their name from the fact that they are designed to help a drafter draw a drawing 'to scale' – draw a drawing in a size that is accurately proportionate to the object, yet in a size that will fit on the paper and be easily read. They come in several different measurements for different types of drawings and several shapes. The most common shape is the triangular shape with six faces. A variation of this is flat scales with four faces. Scales will be discussed in detail later in this book.

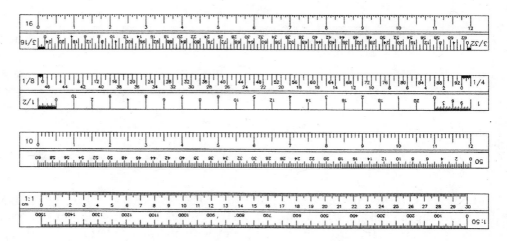

Different types of scales

Scales are made of several different materials - usually of plastic, but some are made of metal or wood. The wooden ones are often faced with plastic. The divisions are engraved into the surface. Scales must be protected to keep the edges and surfaces unmarred. They should be kept clean. Scales are used for measuring existing lines or distances and for measuring distance between two points where a line will be drawn. They are made for measuring only and should never be drawn against or marked on.

Care of Instruments

Drafting instruments are precision aids to the drafter and should be given respect and care. The quality of the drawing will reflect the professional ability of the drafter. Accuracy and neatness are of the utmost importance. Proper care and use of the drafting instruments will do much to aid the drafter in achieving this quality. Misuse can cause instruments to become damaged resulting in inaccurate drawings. With proper use and storage, the instruments will last a lifetime.

T-Squares should be stored lying flat on the board or hanging on a peg by the hole in the end of the blade. Every precaution should be taken to keep the head from being knocked out of alignment. The blade should be kept free of nicks so that it will always give a straight and smooth edge for drawing.

Compasses need to be stored in a case that will protect the point. The point must be true and sharp. The lead should be sharpened for each use. Alignment of lead and point should be checked and adjusted as often as it is used.

Triangles, curves and templates need to be stored flat and protected from scratches and nicks. They should be washed with a mild soap and water occasionally to keep them clean and prevent the transfer of dirt to the drawing.

The protractor should be kept in an envelope or case to keep it clean and to protect the printing for accuracy and ease in reading.

Pencils can be stored in a tray or box to keep the lead from being broken. Of course, leads must be sharpened often – some drafters re-sharpen for nearly every line. This assures consistent line weights.

Pencil sharpeners or pointers should be kept away from the drafting area to prevent graphite dust from getting on the drawing or drawing surface. The sharpeners should be stored in a separate container to prevent the graphite from getting on the instruments.

Technical pens will normally be in a case that protects the points and allows them to stand upright.

The erasing shield is a delicate metal instrument. The smoother it is kept, the better job it will do, so it should be kept from being bent.

The instruments should be properly stored in a cabinet or carrying case when not in use and properly cared for when they are in use. A compartmented box is the best carrying case to use so that the instruments are separated and kept from damaging one another. It is best to select a briefcase or 'tackle box' that will hold all of your instruments, be easy to carry, and allow easy access to the instruments.

TECHNIQUES

There are some techniques that have been found through experience to help the drafter do the job faster, more efficiently, and consistently. Some of these hints could be useful to you.

Setting up the Board

The first step in beginning a drawing is attaching the paper to the board. The sheet is fastened to the board with tape at each corner of the sheet. The sheet should be positioned on the board in a location comfortable for drawing with space to move the instruments around the drawing. The sheet is laid with the edge parallel to the straightedge and tape applied to the top left corner. Holding the paper firmly in place, move the straightedge down and smooth the paper diagonally to the lower right hand corner and apply another piece of tape. Then smooth the paper from the left to right and apply a piece of tape to the upper right corner of the paper. Finally, smooth the paper diagonally to the lower left corner and tape that.

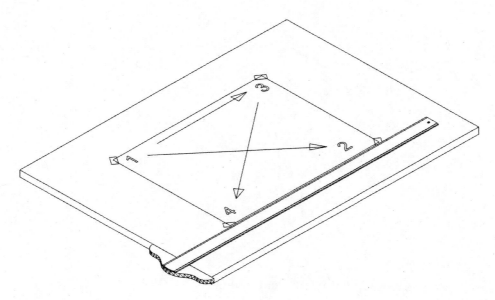

Instrument Layout

When a drawing is to be made, the instruments to be used should be laid out for easy access and use. They should be close at hand and yet out of the way. The illustration below shows how this can be achieved on a table top. In this illustration, the instruments are arranged across the top of the board. They are out of the drawing area, yet within easy reach. If they are always placed in the same location, the drafter will soon become familiar enough to reach for each item without fumbling. If the table is large enough, this is ideal, but if a small board is used, the instruments should be laid out similarly on a table nearby for easy access. Organizing the instruments in a regular order promotes good drawing practice and increases speed of drawing.

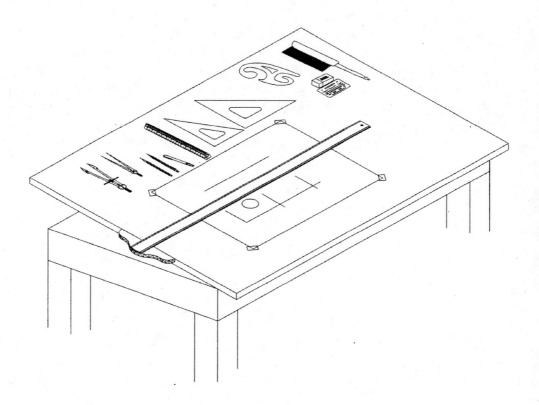

A typical instrument layout on a drafting table

Lettering

Lettering is a very important part of the drawing. Every drawing contains much more than the drawing in order to be complete. Dimensions and notes make the drawing a complete document of information. While the drawing may be clear and understandable, without the dimensions, it is concept only. Dimensions set the parameters and limits. Notes add any needed instruction or explanation that may not be self-evident in the drawing. The lettering of dimensions and notes must be just as clear as the drawing. The lettering becomes part of the drawing and reflects the same quality and accuracy as the drawing. A drafter should take great pride in his/her lettering. Every letter should be *drawn* not 'printed'. We may have learned to print well in elementary school and have developed a fast and personal technique, but for the drafting application, it is necessary to go back to the beginning and re-learn to *draw* the letters. While there may be a few people with natural talent, most need to learn to letter properly. It is a skill which comes with a lot of practice and concentration. It is a skill which needs to be constantly guarded from the tendency to be careless. Select a paragraph from this book or one assigned by your instructor to copy as you practice drawing your letters. Use the hints following to develop your skill.

Usually, lettering on a drawing is all capital letters. Either vertical or inclined lettering is used, but the style should be consistent and used throughout a given sheet and project. Lettering should be consistent in size as well as style.

The following illustration gives examples of the vertical and inclined letters. Some books would instruct you how to make each stroke to form the letters. This book will not dictate that specifically, but with practice you will find that a certain method works for you. Draw straight lines in a single stroke – up or down or horizontally. Curved letters may require several strokes. Find the combination that works best for you. It is often more comfortable for some people to draw the strokes in one direction or another, but the end result should be same. The illustration also shows the proper use of the guide lines.

A B C D E F G H I J K L M
N O P Q R S T U V W X Y Z
a b c d e f g h i j k l m
n o p q r s t u v w x y z
0 1 2 3 4 5 6 7 8 9

A B C D E F G H I J K L M
N O P Q R S T U V W X Y Z
a b c d e f g h i j k l m
n o p q r s t u v w x y z
0 1 2 3 4 5 6 7 8 9

VERTICAL LETTERING   INCLINED LETTERING
A B C D E F G H I J K L M   A B C D E F G H I J K L M
N O P Q R S T U V W X Y Z   N O P Q R S T U V W X Y Z
a b c d e f g h i j k l m   a b c d e f g h i j k l m
n o p q r s t u v w x y z   n o p q r s t u v w x y z
0 1 2 3 4 5 6 7 8 9   0 1 2 3 4 5 6 7 8 9

Guide lines

Lettering guide lines should be used for all lettering. Using the Ames Lettering guide discussed earlier, or other means of creating guide lines, draw very light lines to indicate the space for each letter or word. Then carefully draw each letter or number within these lines. Letters should be consistent in height, width, and angle, as well as line thickness and darkness. Note that both inclined and vertical lettering uses the same guide line spacing. Capitals fill the total space and lower case letters are drawn in the smaller space with the exception of extenders and subtenders of certain lower case letters. The key word is always consistency. Be sure that your lettering is consistent in size, shape, and spacing for a pleasing appearance. Spacing between letters and words is not a mechanical operation, but as is pleasing to the eye. Some letters may be placed closer together than others and give the same appearance of space.

Title Block

Most drawings will require a title block. The title block gives some important information. Depending upon the complexity of the project or drawing, the title block can provide many different types of information, but for all drawings, there are five basic pieces of information:

Title
Date
Scale
Drawn by
Sheet number

The **Title** identifies the drawing(s) on the sheet. It may be one drawing or several on the same sheet so the title must identify accordingly. If only one drawing, the title would be of that drawing, but if there is more than one drawing on a sheet, the title would encompass them all, such as, **'Details'** and each drawing would be titled independently on the sheet.

The **Date** should reflect the date that the sheet was completed. A drawing may be worked on for a few minutes or several days, but the date in the title block should be the day that the drawing was completed.

The **Scale** is the indication of the size of the drawing in relation to the real object. This scale is written in different ways according to the type of drawing that is being drawn. A machine drawing is expressed in a ratio form such as 1:2 or 1/2, which means that the drawing is one half the size of the original or planned object. In Architectural drawings, the scale is expressed in a relationship of measurements, such as 1/4"=1'-0", which means that the measurement of 1/4" on the drawing represents a measurement of 1' on the actual object. In Civil drawings, the scale would also represent a relationship similar to the architectural, but in a larger scale and relates an inch to a greater distance, such as 1"=30'.

Drawn by identifies the drafter – his/her name or initials - so that if there is any question about the drawing, it is possible to find the one who did the drawing.

Sheet Number identifies the drawing much as page numbers in a book. Since there may be more than one drawing on a sheet, it is the sheet that needs to be numbered more than the drawing. In a set of drawings, this would be numbered as the sheet number in relation to the total number of sheets in a set, for example, *sheet 1 of 6*.

Lettering in the title block is dictated by space and design of the title block. Borders and title blocks usually become a standard of the company or designer. Once that standard is set, each drawing would use the same title block and lettering style and size. Normally, the title of the sheet is larger so that it will stand out for immediate attention. The other blocks within the title block can be and will be arranged as seen fit by the designer or firm. The important factor is the appearance and information presented. The title block should be easy to find and read. All title blocks should be in the same location on the sheets in a set, and all information within the title block should be consistent. The title block is the introduction and information index to the sheet. The reader of the sheet will start with the title block for the information he/she needs to understand what to expect on the sheet.

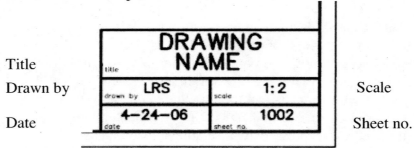

The above title block has only the basic information, but many other sections can be added to a block to reflect the company name and logo, a section to show who checked or approved the drawing, an area to indicate when revisions were made to the drawing, etc. Additional blocks could be put on the drawing to show materials and processes.

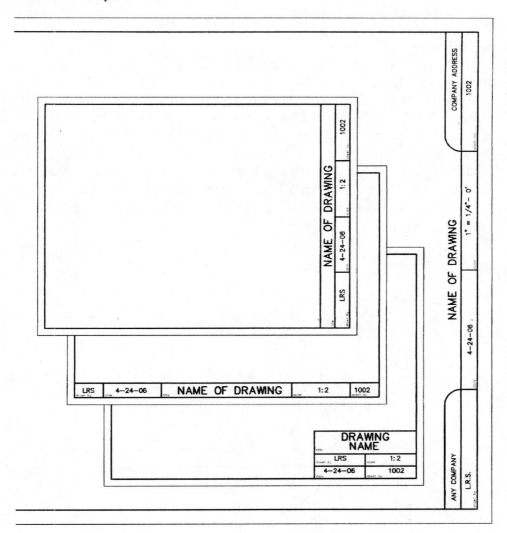

Above are samples of basic Title blocks. Each gives the same information, but with different configurations. Notice that some are horizontal and some are vertical on the sheet. The drawing should be approached and read from the bottom or right end of the sheet, so the title block should be orientated with that consideration. On large architectural sheets it is common to use the end of sheet vertical title block so it can be easily read from a roll of many sheets.

Drawing Lines

Every drawing is simply the combination of lines – vertical, horizontal, angled, and arcs or circles. Using the information given and the knowledge of how to use the instruments is combined to create a 'picture' of a thought with lines.

Lines should be drawn by pulling the pencil across the paper along the straightedge or triangle. Horizontal lines are drawn from left to right along the straightedge. Vertical lines are drawn from bottom to top, angled lines as shown in the illustration below, using triangles as the straightedge. The pencil should be rotated as it is pulled along for even width of line. Consistent pressure assures an even darkness of line. Line thickness is best achieved by different sharpness of lead. Lines should be drawn accurately and meet precisely at intersections with no overhanging ends or space between ends. Practice drawing lines horizontally and vertically, and at various angles.

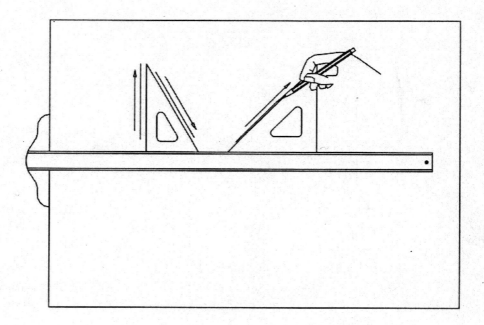

Geometric Techniques

Much of what is done with lines in drafting involves some understanding of the basic principles of geometry and the geometric construction of lines. Some basic techniques follow:

Divide a line into equal parts –

To divide line **AB**, draw a vertical line down from point **B**. Use a scale or ruler to find the desired number of spaces on it. Line the **0** up with point **A**. Rotate the ruler until the desired division number meets the vertical line. Draw vertical lines up from each number to the horizontal line. The result will be an evenly divided line. The example shows a line divided into eight parts, but any number can be used. The total distance of the divisions on the ruler must be slightly longer than the line to be divided. The angle can be set at any angle between the **A** and the vertical line from the **B**.

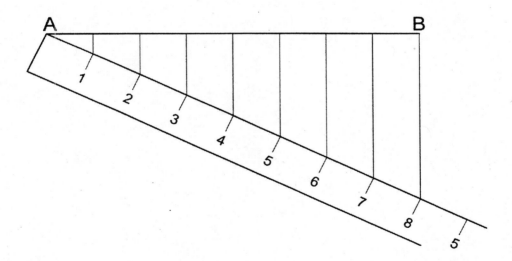

Any line can be divided into two equal parts using a compass. Set your compass to any setting that is more than one half the length of the line. Using point **A** & **B**, strike arcs to create points **C** & **D**. A line between **C** & **D** will divide (or bisect) the line into two halves.

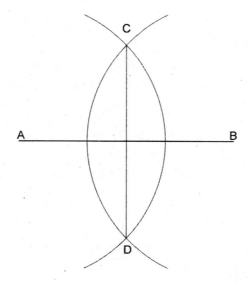

Bisecting a line

The same procedure can be used to bisect an arc as for a line.

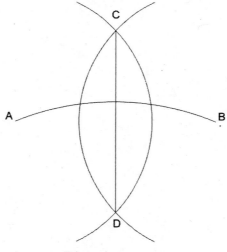

Bisecting an arc

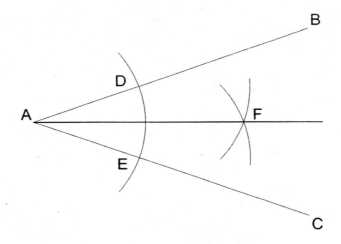

Bisecting an Angle

To bisect angle **ABC**, follow the steps illustrated above. Using **A** as a center point, strike an arc at any distance to create points **D & E**. Using **D & E** as centers, strike arcs to create **F**. Draw a line from **A** through **F** to divide the angle in half.

Create a triangle with given side lengths.

If you know the lengths of the triangle but not the angles, as above where three lines are given, it can be easily constructed with a compass. Draw line **AB**. Set your compass at the length of line **cd**. Using point **A** as the center, strike an arc to **Y**. Repeat the process, using the radius of the length of line **ef** and strike an arc using point **B** as the center. The intersection at **Y** will be the apex of the triangle. Lines drawn between points **AB & Y** will form the triangle.

To find the center of a circle when the center is not known, draw two lines as in the examples below to form chords on the edge of the circle as **ab** and **ac**. Bisect these lines. The point where the two bisecting lines meet will be the center of the circle or arc.

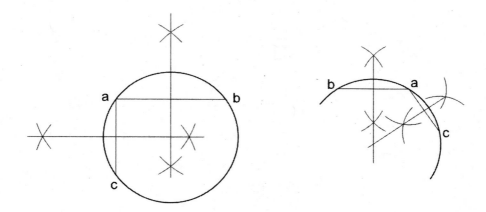

Finding the unknown center of a circle or arc

Connect two circles or arcs with a tangent arc

Set your compass to the distance of the radius of circle **1** plus the radius of the desired joining arc. Using the center of the circle, draw arc **ab**. Repeat for circle **2**. Set the compass to distance of the radius of circle **2** plus the radius of the joining circle. Draw arc **cd**. Set your compass to the radius of the joining arc. Using the intersection **Y** at the intersection of **ab** and **cd** as the center point, draw the joining arc.

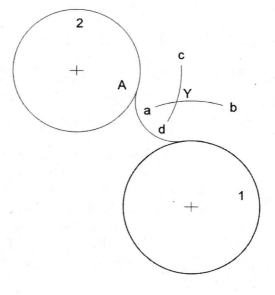

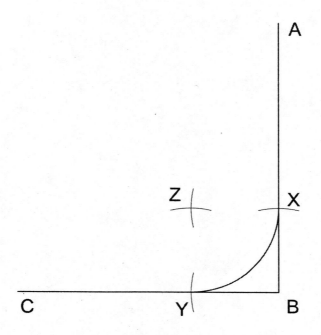

Connect two straight lines with an arc.

When two straight lines are to be connected with an arc, the procedure is as illustrated above. Set you compass to the radius of the desired arc. Using **B** as the center, strike an arc on each line as **X** and **Y** above. Then, using **X** and **Y** as centers, and keeping the same compass setting, strike the two arcs to create point **Z**. Still keeping the same compass setting, use point **Z** to create the desired arc between points **Y** and **X** , joining the lines with a tangent arc.

ORTHOGRAPHIC PROJECTION

ORTHOGRAPHIC PROJECTION

Orthographic projection is a system of viewing objects and transferring them to paper in true size and proportion. When an object is viewed with the normal eye, we see it from several sides. We see it in relationship to other objects. We see it distorted in a way, because our eyes do not measure actual dimensions. We see the object and our mind overlooks the fact that distance and angle distort the image in our eyes. Our mind measures perceived dimensions.

To overcome this distortion for the drawing process, we separate the object into 'views' and create 'working drawings'. We separate and place these views in relationship with one another. This is called **Multi-view** drawings. It is the process of looking at an object one side at a time and then placing these sides together on one sheet. For this we use Orthographic Projection. There are two common types of Orthographic Projection:

First Angle projection and **Third Angle projection**.

> **First Angle projection** looks at the object and projects past the object to 'trace' the lines of the object on a plane behind the object. First Angle projection is used Europe.

> In the United States, we use **Third Angle Projection**. Third Angle projection views the object and 'traces' the outline of the object on a plane in front of the object. Each side of the object is looked at directly and the lines that are seen are transferred to a plane called the Picture Plane in front of the object.

To best explain the process of Orthographic projection, an imaginary glass box is used. The object 'floats' inside this box to be equal distance from its sides. When viewed, the glass sides act as a drawing plane and the view that is seen through the box is traced on these planes. See the illustrations on the following pages to understand this process.

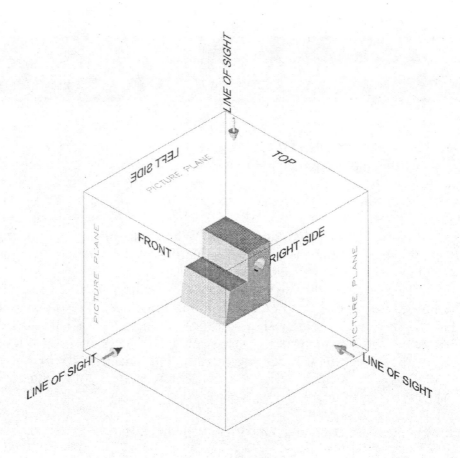

The Imaginary Glass box

The object is placed in an imaginary glass box and each side of that object is viewed from a direction perpendicular to that side.

When the object is placed in the glass box, the most descriptive side, or the side most commonly thought of as the 'front' is usually chosen as the front and the other sides are identified accordingly. Imaginary lines called **projectors** are projected from the object to the glass in front of it (between the object and the viewer) to trace the object outlines on the glass or plane called the **viewing plane**. All lines on a given view are parallel to the plane that the view is projected to.

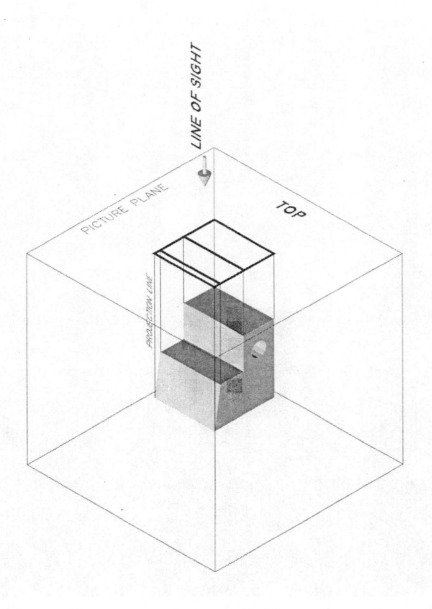

Top view projected from top of object to the Picture Plane

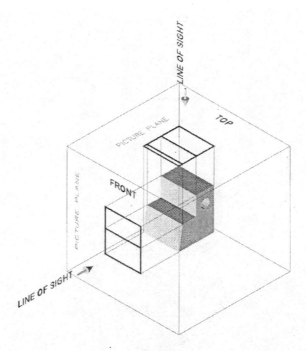

The Top and Front projected to the picture planes

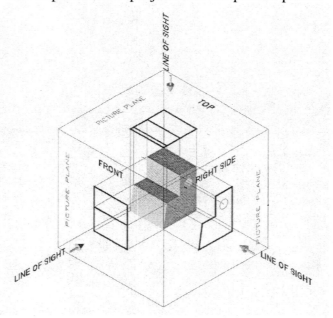

Three sides projected to the picture planes

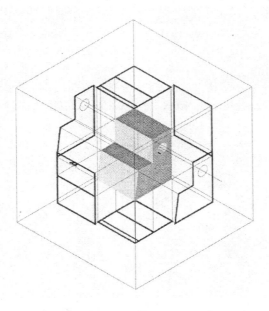

This is repeated for all six sides of the glass box. When the drawings are projected and drawn on the glass box, the box is opened or folded out. The edges where the box is hinged are called **folding lines**.

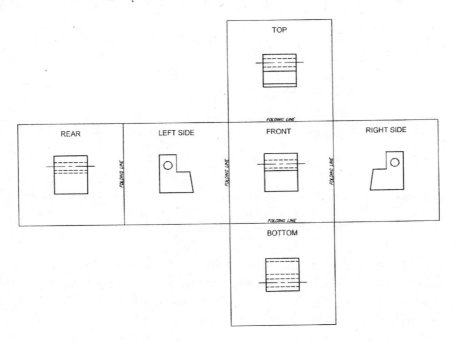

The folded out sides are kept in relationship to one another and become the six basic views of multi-view drawings. The spacing between views may be adjusted for space on the drawing sheet, but the relationship remains constant. This relationship is shown in the folded out illustration. The most commonly used views are the front view, top view, and right side view. The other views may be used for clarity when necessary.

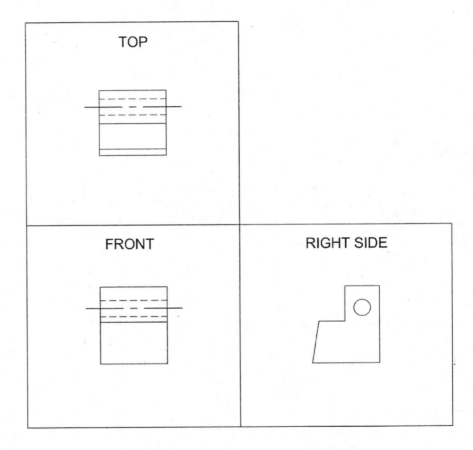

The three most common views

Remember that the folding lines are only imaginary and do not need to be drawn between views, but the views should always maintain the same relationship. When properly placed, the views relate to each other and edges become lines which can be transferred from view to view to aid in the drawing process. Lines project from left to right and from top to front or front to top to transfer the measurements. Usually the most descriptive view is drawn first and information is transferred using these projection lines.

Also, information can be transferred with the 'magic 45'. As illustrated in the following diagram, a 45° angle line placed at the intersection of imaginary folding lines can be used to project lines from the top view to the side view or in reverse from the side to the top.

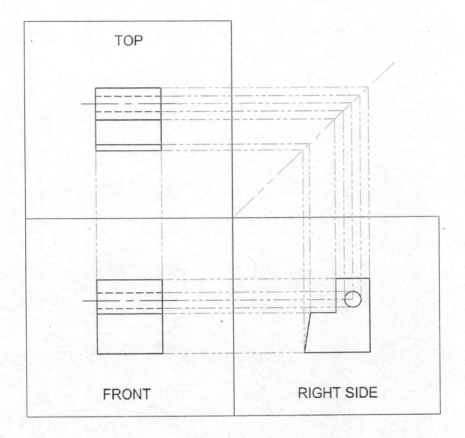

Whether drawing in the Mechanical, Electrical, Civil, or Architectural field, knowledge of orthographic projection and the placement of views is the base for a good understanding of drawing principles. All areas of drafting will use this concept to produce detail drawings.

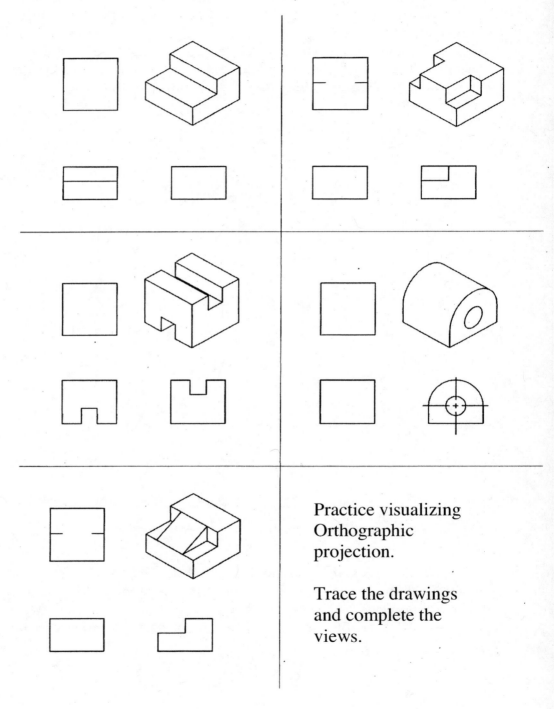

Practice visualizing Orthographic projection.

Trace the drawings and complete the views.

GETTING STARTED

When starting a drawing, there are beginning steps that should be followed to develop the drawing properly with the desired final results.

Planning the drawing.

A drawing should be placed on the sheet in a pleasing presentation. It must also be planned for proper communication. To do this the drawing must be laid out according to orthographic layout and dimensioning rules. Great care should be taken in laying out the drawing in the beginning so that it will achieve the best results in the end.

Consider the object to be drawn. Look at its size, shape, and features to decide which views to use.

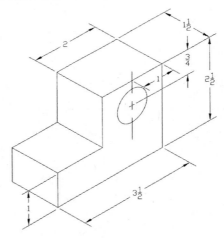

Laying out the drawing

When the views to be used have been decided on, it is time to lay out the drawing and begin. The first step is to select the paper and attach it to the board. This was discussed in detail earlier.

Once the paper is in place, the drafter needs to plan the space needed for the drawing. Select the view(s) and appropriate the space needed for the drawing and other information that will be put on the sheet.

Using the understanding that you have of orthographic projection, you can envision the relationship of the views.

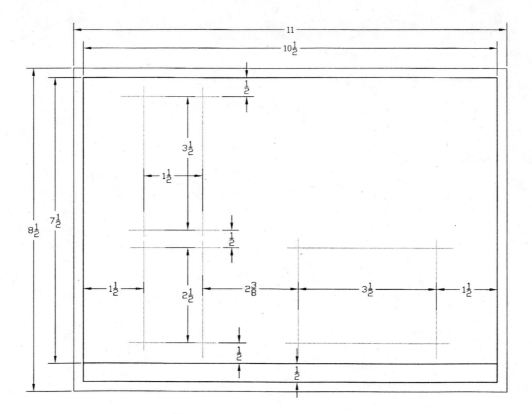

Looking at the overall measurements of the object to be drawn, you can calculate the amount of space you need and subtract it from the space available, then distribute that as space between and around the views. At first this will seem like a complicated mathematical problem, but with a little practice, you will find that the numbers can be rounded off and the space allocated to create a pleasing appearance to your drawing.

Using light construction lines, the drafter will block in the space needed for each view, allowing room for dimensions and notes to be added later. The finished drawing should give the appearance of being centered in the sheet. This may not mean that it is actually 'centered' on the sheet, but that the arrangement of drawings and notes are spaced evenly on the sheet with good margins and space around them. The drawing should not give the appearance of being crowded; nor should the views be separated so far apart that they appear unrelated. Spacing requires some basic math and an eye for spacing. The blocked in areas should give a good indication of the final appearance of the drawing.

With the blocked in areas located, the actual drawing can be started. All drawing should be done lightly first. It is easier to erase and make changes to light lines than to finished dark lines.

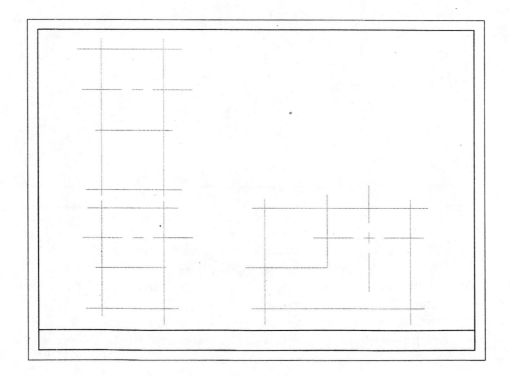

Draw center lines first. If the drawing has circles, arcs, or symmetrical areas, it will require centerlines. These should be located and drawn accurately first and the rest of the drawing relate to them as appropriate. Centerlines must be thin, sharp lines to give the accuracy that is needed when using the compass. Centerlines should cross at the exact point of the center of the arc or circle.

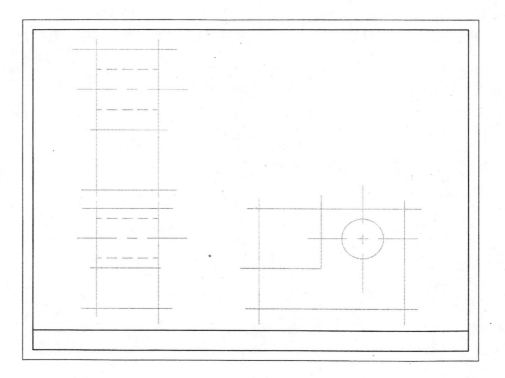

Draw Circles and arcs. Set the compass to the desired setting and place the point exactly on the cross of the centerline. Be careful to set the point correctly the first time. If it is misplaced and reset, the hole that is made in the paper may become enlarged or tear the paper. It will then be difficult to keep an accurate location. Strike the arc or circle by rotating the compass around the center point.

Draw horizontal and vertical lines next. Lines should be sharp and meet exactly at every intersection. Strive for accuracy in the beginning stages of the drawing. Accuracy will then continue throughout the entire drawing.

Hidden lines are added to show features that would not be visible on that view.

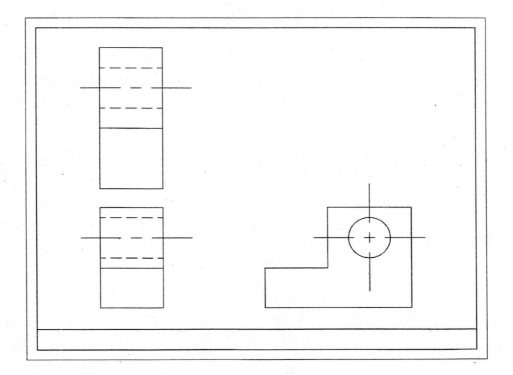

Darken all lines. As you darken lines, you should work from the top of the sheet down so you are not working over the top of lines and smudging them. A piece of paper may be placed under your hands and instruments to protect your work from smudging.

Dimensions and notes would be added next. This will be covered in another chapter.

The Title Block will be filled in as the last step. This is kept until last to have all the final data, date and scale when the drawing is finished. It also prevents smearing the lettering in the title block during the drawing process.

DRAFTING CONVENTIONS

Alphabet of Lines

Lines on the drawing speak a language of their own. To know the meaning of each line type or thickness is another method of conveying the message of the drawing. The different line types represent the outline of the object, location of circles, hidden features, indication of cut lines, as well as dimension and note locations. The line thickness aids in this information. The following 'Alphabet of lines' is a standard representation of some of the lines most used by drafters and understood by print readers.

Object line	————————————————	Thick
Border line	————————————————	Extra Thick
Center line	— — — — — — — —	Thin
Hidden line	— — — — — — — — — —	Medium
Phantom line	— — — — — — — —	Thin
Dimension Line	|←——————10¾——————→|	Thin
Cutting plane Line		Extra Thick

All lines should be sharp and black regardless of thickness. Lines should be consistent and used for the proper purpose consistently. The uses of each line are defined as follows:

Object Lines are continuous thick lines that are drawn as the edges of the object itself. They should completely identify the outline of the object and each solid part of the object.

Border Lines are extra thick continuous lines and are used to show the edge of the drawing space – giving the desired margins - and also outline of the title block. They are extra thick to present a bold border and separate themselves from the drawing.

Center Lines show symmetry. They are thin alternating long and short dashed lines used most commonly to show the center of arcs and circles, but may also be used to show the center of any symmetrical object.

Hidden Lines show features that are not visible on the surface of the object. They are medium in thickness. Although they may appear confusing if there are many interior or 'hidden' features, it is important to show the whole object so hidden lines are dashed and thinner than object lines to offer separate information.

Phantom Lines are unique lines that show alternate features of an object or part of an object. They are thin lines drawn with long dashes and two short dashes.

Dimension Lines are thin continuous lines that carry the dimension. Dimension lines extend between extension lines and are terminated with an arrow head at each end.

Cutting Plane Lines show the cut through an object for section views. The Cutting Plane line is drawn extra thick and has arrow heads on each end to show the direction of viewing the section.

Although there are other means of communicating information, the proper use of lines makes the drawing more than just lines. It gives to the print reader an instant recognition of the object that the drafter is trying to portray.

Symbols

Another method of communication is the use of symbols. Symbols give an instant recognition of a message - like road signs where a diamond shape means caution, octagon means stop, etc. Symbols on a drawing are not as much shapes but are representation of materials/fixtures/equipment that would be difficult to show on the drawing in another way. The use of these symbols is dictated by the type of drawing. Some of these symbols are shown in the appendix.

MEASURING

MEASURING

Measuring is the beginning of most drawing projects. Measurements for the object may be given to you, the drafter, by another person, or you may obtain them yourself. At this point, however, you are most interested in drawing the working drawings accurately from the measurements given.

Although there are several different scales to be considered, the scale is basically a measuring device. The drafter uses a scale to measure and locate all features on a drawing. It is used much as one would use a ruler. To lay out a distance on your paper, draw a light line in the direction that you want to measure. Lay the scale flat on the paper along that line with '0' at the beginning point. At the distance that you want to mark off, use a sharp pencil and make a dot beside the scale on the paper precisely at the right distance. Repeat this process for any distances you wish to mark. Using your triangle or straightedge, draw a line perpendicular to the marked off line and dot. Never draw along the edge of the scale.

Transferring measurements may be done with dividers. If you want to repeat a distance, you may set your dividers at two points on a line and then place them along another line, making small prick points in the paper to mark that distance. This can be repeated again and again as long as the setting is kept true on the dividers.

Engineer's Scale

The engineer's scale is so named because it is used primarily for engineering projects where measurements are in inches and tenths of inches, or feet and tenths of feet. It is used by civil engineers and by mechanical engineers who measure in decimal measurements rather than fractions. One face of the engineer's scale reflects 'full size' – inches and tenths of inches. The other faces reflect distances that are proportionate to the inch and tenths.

When used by engineering drafters, they can be read as feet and tenths of feet – for example, a drawing that represents 1 foot could use any of the faces and draw a drawing that represented 1 foot as one inch, two inches, three inches, four inches, five inches, etc. on the paper. The divisions between the numbers would represent one tenth of a foot.

They can also be used to draw larger distances representing the same proportion in feet - for example, a drawing representing 10 foot, could use the same scales and represent 10 foot as one inch, two inches, three inches, etc. The different scales make it easy to set the scale without calculations for each distance. A drawing that represents twenty feet on a sheet could be drawn with the 20 scale and each inch would represent 20 feet. The divisions between the numbers would be feet. The scale on the drawing would read 1" = 20".

The engineer's scale is divided into inches and tenths of inches. It can be used to measure inches or in proportionate scales as shown. Since the divisions are in increments of tens, many combinations can be used.

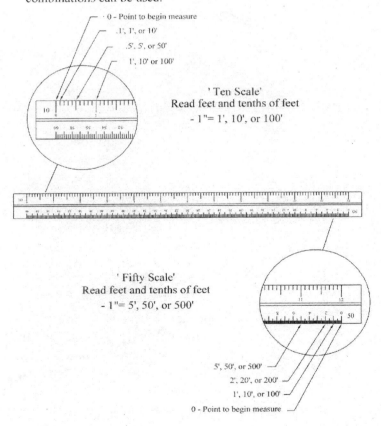

· 0 - Point to begin measure

.1', 1', or 10'

.5', 5', or 50'

1', 10' or 100'

' Ten Scale'
Read feet and tenths of feet
- 1 "= 1', 10', or 100'

' Fifty Scale'
Read feet and tenths of feet
- 1"= 5', 50', or 500'

5', 50', or 500'

2', 20', or 200'

1', 10', or 100'

0 - Point to begin measure

Metric scale

The metric scale is based on divisions of a meter. A meter is divided into centimeters (cm) – 100 cm = 1 meter , and millimeters (mm) 10 mm = 1 cm. Scales are set to be reduced proportionately to these divisions – by ratio, Drawings that are done in metric scales are done in a desired proportion in relation to full size – full scale, half scale, or greater scales, for example – 1:20, 1:40, etc.

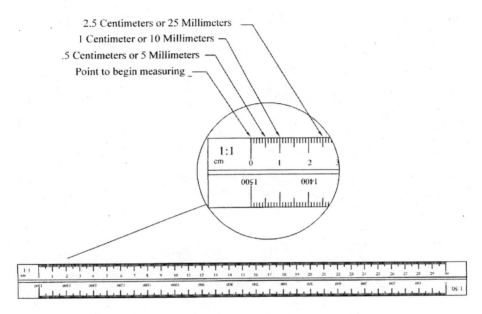

The Metric Scale

Architect's scale

The architect's scale refers to feet, inches, and fractions of inches. The full size face is a 12" measure divided into inches and fractions of inches – as small as 16ths or 32nds of an inch. This is used like any ruler. On a triangular scale, there are five other faces. These faces represent 'scales' or size representations for different proportions. Each face has divisions starting on either end with a total of ten different scales. On each end of the five faces there are marked 3/32, 3/16, 1/8, etc. The ten different scales give different options from 3/16" = 1'-0" to 3"= 1'-0". Each scale is set with pre-calculated divisions to make it easy to draw a drawing at the chosen scale. The starting place for that division will be marked by a '0' and will be read away from the '0' in graduations appropriate for that division. The division tells the distance that is used to represent a foot on that scale. For example, 1/8 means that each 1/8th of an inch represents a foot. Each 3/16th of an inch represents a foot and so on. To the left of the "0", the division is divided into 'inches' to that scale. A measurement may be made in feet and inches to any of the ten options of scale.

The shown examples describe how each scale is read.

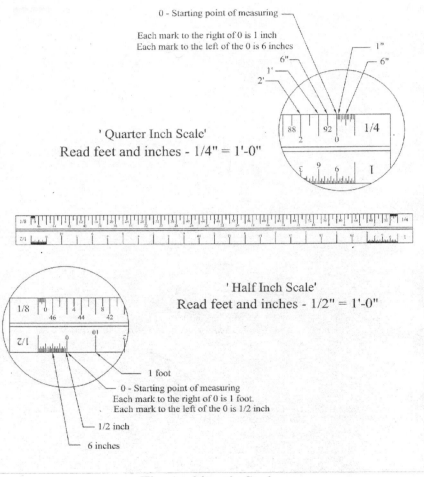

' Quarter Inch Scale'
Read feet and inches - 1/4" = 1'-0"

0 - Starting point of measuring

Each mark to the right of 0 is 1 inch
Each mark to the left of the 0 is 6 inches

' Half Inch Scale'
Read feet and inches - 1/2" = 1'-0"

1 foot
0 - Starting point of measuring
Each mark to the right of 0 is 1 foot.
Each mark to the left of the 0 is 1/2 inch

1/2 inch

6 inches

The Architect's Scale

To measure and draw a distance in feet and inches, put the '0' on the beginning point and measure in feet away from the '0'. Mark the desired foot point; then slide the scale past that mark by reading the inches to the left of the '0' the desired number of inches and make a second mark at the new location of the foot mark. The second mark will represent both feet and inches.

Use the lines below to practice your measuring skills.

a. ————————————

b. ——————————————

c. ————————————————

d. ——————————————————

e. ————————————————————

f. ————————————————————————

g. ——————————————————————————

h. ————————————————————————————

i. ——————————————————————————————

j. ——————————————————————————————————

Measure each line with various methods. First measure each one in inches and fractions to the nearest 1/16 inch. Then measure each line with a metric scale, measuring in mm and cm. Then measure to scale with selected scales of 1'=10', 1"=20', etc. Measure each line using the architect's scale with selected scales of 1/4" = 1', 1/2" = 1', etc. Measure to the nearest inch - to scale. On a sheet of paper, record your measurements for each line as you measure and compare the results.

DIMENSIONING

DIMENSIONING

Dimensioning is a very large part of drafting. Even the best and most detailed drawing will show only shape and proportion. The manufacturer still needs the 'facts'. Dimensions are the 'facts'. These are the recorded measurements necessary to 'build' the object. Whether a machine part or a huge building, the manufacturer cannot complete the construction without all of the dimensions. The drafter must take care to be sure every dimension is included that is necessary to manufacture the object. Dimensioning methods vary according to the type of drawing, but for the purpose of this book, the very basic methods will be discussed. There are basic rules that dictate what should be dimensioned. There are also basic rules that dictate the way that the dimensions should be put on the drawing.

Dimensions are divided into two basic categories - **Size** and **Location**. As the titles imply, Size Dimensions relate to measurements of a part. Location Dimensions relate to where a feature is located on the part. In other words, the diameter of a hole drilled through a part is the size. The distance from the edge of the part to the hole is its location. The overall distance from one end of the object to the other relates to size. The distance from the edge of the object to an offset in the object is its location. The two categories of dimensions work together to give all the facts necessary for the manufacture of the part.

Although dimensioning is a very basic part of the drawing, it is a very important part. Just as important as the existence of dimensions, it is important that the dimensioning be done correctly. Dimensions must be:

Complete – enough information to manufacture the object
Readable – easily understood
Usable – require little confusion or calculation by the reader
Accurate – precise size and location measurements
Dimensions include four parts:

Extension lines – lines perpendicular to the object, and extending from the object to indicate the limits of the dimension to be expressed.

Dimension lines – lines parallel to the object which carries the dimension. Dimension lines extend from one extension line to another.

Arrow heads – arrow heads end the dimension lines and point to the extension lines. Arrow heads should be drawn at a proportion of 3:1 – three times as long as they are wide. The size of the arrow head is proportionate to the drawing and space. All arrow heads on a given drawing should be the same size.

Dimension – dimensions are the printed text of the distance dimensioned. Dimension text should be consistent in size and style throughout the drawing.

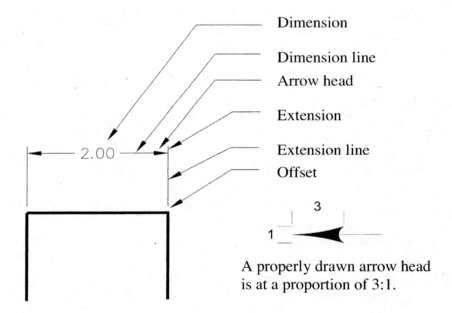

A properly drawn arrow head is at a proportion of 3:1.

The above illustration defines the parts of the dimension. Notice two terms not included in the 'four parts' of the dimension – **Extension** and **Offset**. While they are not 'parts' of the dimension they are part of how a dimension is placed on the drawing – The extension is an approximate 1/16" distance that the extension line projects past the dimension line. The Offset is an approximate 1/16" distance between the object and the beginning of the extension line. These distances help to prevent confusion in the dimension. The offset keeps the extension line from being confused with the object lines. The extension gives a clearer image of the relationship of the dimension line and extension line.

There are two types of dimensions – **Unidirectional** and **Aligned**. Unidirectional dimensions are all read from one direction. This is the most common practice with machine drafting. Aligned dimensions are lettered parallel to the dimension line. This is sometimes used in machine drafting, but more commonly in Architectural drafting.

Dimensions should be placed on the most descriptive views – one or all views that show features of the object. In the below example, the dimensions are shown giving both size and location of the features.

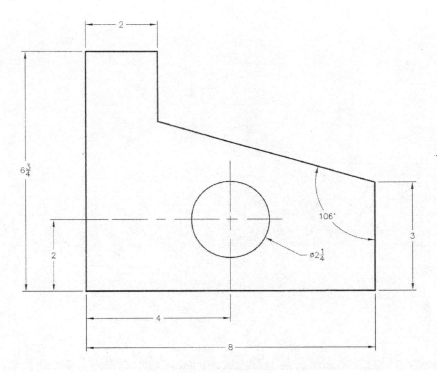

Fractional Dimensions

Note that the dimensions are placed outside the view as much as possible. They are evenly spaced. The small dimensions are placed first and the larger dimensions outside of the small ones. The dimensions are placed in a break in the dimension lines. All the dimensions read from the bottom of the sheet. This is an illustration of unidirectional dimensioning.

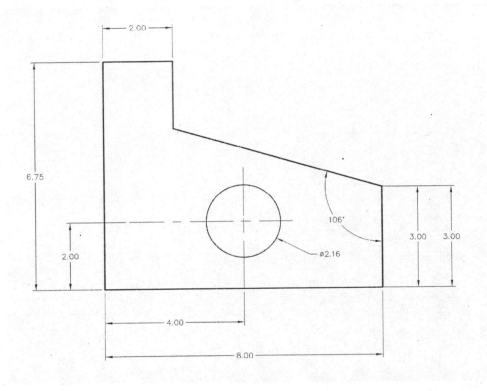

Decimal Dimensions

The following illustration shows the method of dimensioning circles and arcs. Circles are dimensioned in Diameter measurements, Arcs are dimensioned in Radius measurements. These dimensions are best carried on the outside of the object on a LEADER. The leader is an arrow that points to the center of the circle if placed outside and from the center to the arc if placed inside. Notice the use of the R for radius and the circle with the slash through it is the symbol for diameter. Diameter can also be expressed with a D or DIA.

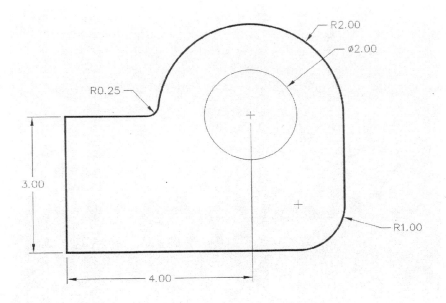

Dimensioning arcs and circles

Dimensioning Architectural drawings uses slightly different parameters. Dimension lines and extension lines remain the same, but the layout is different. The dimensions are placed 'above' the dimension line, and are Aligned dimensions, read from the bottom of the drawing or from the right end. Dimensions are given in feet and inches, with a common practice to show both feet and inches even if the number would be a '0'. This prevents any question of whether one was mistakenly omitted. There is no break in the dimension line as with machine drawing. Architectural dimensions also are 'stacked' - giving total short dimensions, then longer dimensions in the next stack, and longer dimensions in the next stack. This is repeated for each feature.

The below illustration shows dimensions in the first stack on top giving dimensions from the outside of the building to the centers of window, partition, window, and then to the outside of the outer wall on the other end. The second stack shows the dimensions from outside wall to center of partition, to outside wall. And then the third stack gives an overall dimension. The same practice is done on the end of the building and at the bottom. All stacks are carried all the way through so that the reader does not need to calculate any part of the distance. Some dimensions that would be repetitive such as overall dimensions top and bottom are not repeated. This drawing shows traditional arrow heads, but some prefer to substitute the arrow head with slashes or dots.

All dimensions are carried outside the drawing as much as possible, only dimensioning inside the drawing when absolutely necessary.

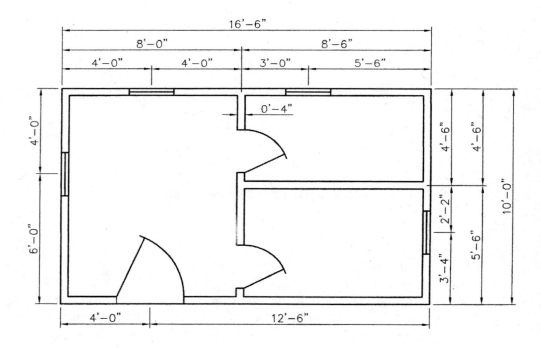

Architectural Dimensions

SECTIONING

SECTIONING

Section drawing is a method of showing materials and the interior of parts. There are several methods of doing this and each has its own type of section view. It is most commonly used in machine drafting, but also in other types of drafting. The location of the cut is indicated by a ***Cutting Plane Line***. The cutting plane line shows the direction of view with arrow heads at each end.

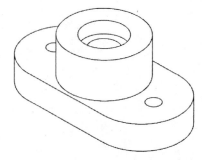

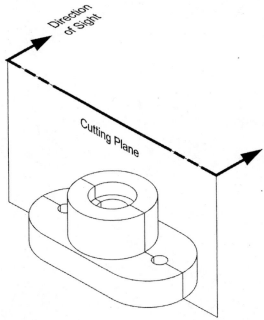

The cutting plane is an imaginary plane that passes through the object wherever it is necessary to show the inner features that the drafter wants to give more definition to.

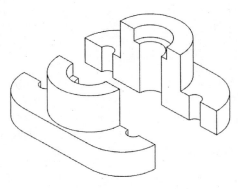

With the cut made, the front section is removed to expose the interior and the cut material.

The result is a view of a partial object that allows the viewer to see how the object is made, or better show interior features. The cut areas of the sectioned view are indicated by **Cross Hatching**. Cross hatching is drawn with thin lines typically 1/16th of an inch apart at 45°.

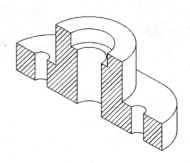

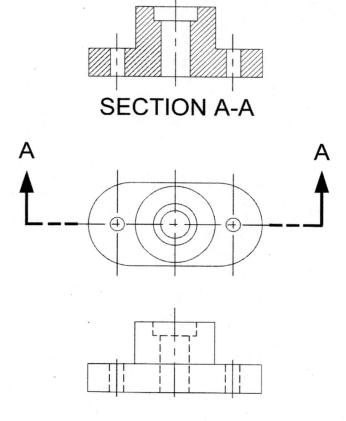

SECTION A-A

A typical drawing of an object, showing the front view and top view with the cutting plane line, and the sectioned view. In most cases, section views are additional to the basic views. The section view is referenced to that view by the use of letters at each end of the cutting plane line which are repeated as part of the title of the section view.

There are many types of section views for different circumstances. Below are listed each and its use.

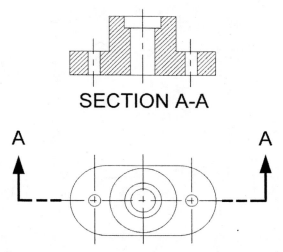

SECTION A-A

Full section

The cut is cut through the entire object, cutting it in half. The title infers that it is a sectional view of the full object.

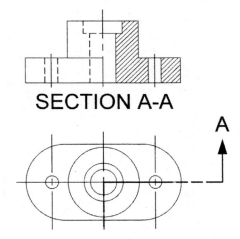

SECTION A-A

Half section

A half section is similar to the full section, except it only cuts half way through the object, exposing one quarter of the inside of the object.

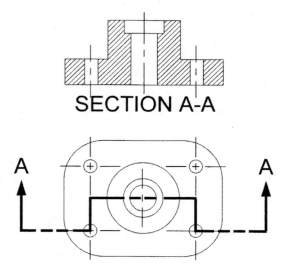

SECTION A-A

Offset section

An offset section is similar to the full section, except the cutting plane line takes a path other than straight through the object. This is done to show interior features that may be offset from, a straight path. The section view gives the appearance of being a straight path through the object.

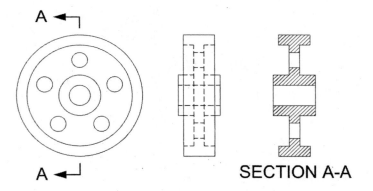

SECTION A-A

Aligned section

An aligned section is most often thought of as a section through a wheel with uneven features such as the holes in the illustration or for spokes. The section aligns the holes as though they were opposite each other (or aligned). Aligned sections may be used for other cases when the section would best be served by aligning a part of the object. In most cases, ribs or spokes are not cross hatched in this type of section.

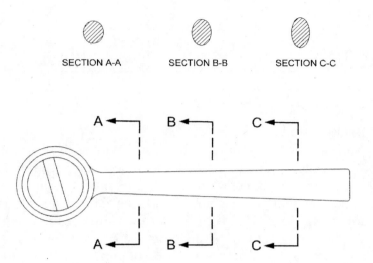

SECTION A-A   SECTION B-B   SECTION C-C

Removed section

A removed section makes a cut through a part as a slice, takes that slice, revolves it 90° and moves it away from the object, usually showing it nearby, as above. This is usually done to show the shape of an object as in this wrench handle.

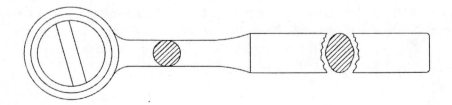

Revolved section

A revolved section is similar to the removed section except when the cut is made it is revolved in place and either shown within the object as on the left above or in a break in the object as on the right above.

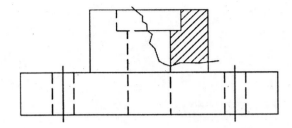

Broken out section

A broken out section reveals a small part of the interior of
an object as though part of the object had been broken
away. This is done when the whole inner feature is not
important, but some clarity would be gained by allowing
some exposure. This type of section view does not require
another view but can be done on one of the main views. It
does not require any identification of cutting plane or title.

Section views are not usually dimensioned. Hidden lines are not
usually shown in sectioned views. As was mentioned earlier, the exposed
material is 'cross-hatched' with thin lines, usually at 45° and spaced
evenly across the cut material. The spacing for the cross-hatching is
usually thought to be about 1/16" apart, but consideration should be given
to the size of the drawing which may dictate that the lines be closer or
allow that they be spaced farther apart to give a more pleasing drawing..
While the 45° lines are the generic symbol, other symbols may be used to
indicate different materials. See the appendix for some of these symbols.

In the case of two or more mating parts in a section as in the drawing below, the hatching should be drawn in different directions to show that the parts are indeed different parts.

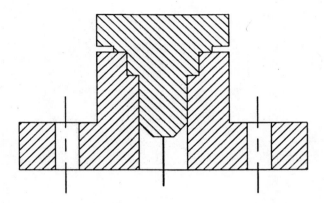

A sectioned view of two mating parts

On the next page are eight sets of views with cutting plane lines. Use these drawings to practice visualizing section views. On a piece of vellum, trace the drawings and turn them into a sectioned view.

Remember to change hidden lines to solid where they become visible exposed edges in the sectioned view. Use hatch lines to show material that has been cut. Keep center lines.

See if any other sections could be used in place of those indicated. Choose which views could be shown as full or half sections. Try broken out sections. Look for an opportunity to use a revolved or removed section.

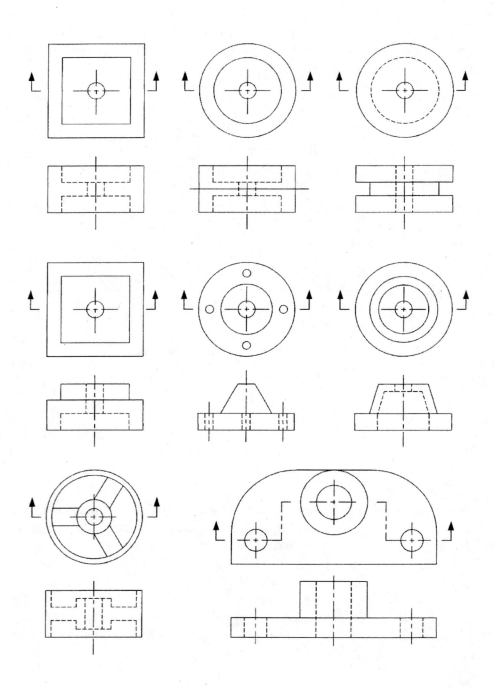

Sectioning Exercises

PICTORIAL DRAWINGS

PICTORAL DRAWINGS

Pictorial drawings aid in visibility of an object. As a drawing is viewed in orthographic views, it is seen as parts for manufacture and assembly. These are 'working drawings' The pictorial drawing depicts the final product more as the eye would see it. Although it is drawn on a flat surface (2-D), it appears to be three dimensional.

Perspective Drawings

Perspective drawings most duplicate what the natural eye sees. As an object is moved farther away from the eye, it appears to become smaller. For example, If you stand beside an auto, it is 'full size' and easily large enough to sit in, but as it drives away, it appears to grow smaller and smaller. While the mind knows the actual size, the eye sees it shrink to a fraction of an inch in size. This illusion is drawn on paper in perspective - a drawing which employs the use of 'receding lines' to 'vanishing points' on the horizon. The drawing would be impossible to measure, but it 'looks' natural. Two types of perspective drawings are **One Point** and **Two Point** perspectives.

> **One Point perspective** directs all lines to one vanishing point. It is as though the eye is looking at one point and everything that is in the eye's view recedes to that point. Such as looking down a railroad track and seeing the rails 'come closer together' in the distance.

> **Two point perspectives** direct all lines to two different vanishing points in opposite directions. The eye is looking at a central point and the view is divided to recede in two directions away from it.

The process of drawing even a small perspective is complex and time consuming. Although this book will not detail the methods of perspective drawing, it will show the concept. The following examples will give an idea of the appearance of perspective drawings.

Below is an example of one point perspective, and a sketch of how it was accomplished.

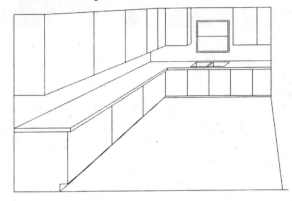

A One Point Perspective

This drawing of a kitchen gives the Appearance of the Cabinets growing Smaller as they are farther away from you. All lines are directed to one vanishing point.

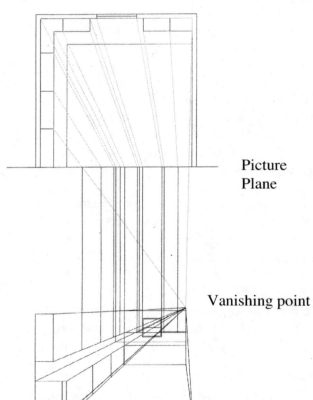

Picture
Plane

Vanishing point

Developing a One Point Perspective Drawing

In this drawing, you can see that the plan view was used. Lines were directed toward the vanishing point onto a picture plane and transferred to lines directed to the vanishing point from the 'front' view.

One Point Perspectives

Two point perspectives give the same 'reality' to the drawing, but use two vanishing points. The two point perspective shows three sides of the object and can be viewed from many levels to see from above, or below or at a natural eye level.

The two point perspective of the small greenhouse below is viewed from a slightly higher level than the natural eye level and illustrates how the lines recede to vanishing points in two directions. Notice that the farther the drawing goes toward the left and right, the smaller everything becomes, not only in vertical dimension, but also in horizontal dimension. The vertical distances are measured on the front corner and recede from that in both directions. This is a mechanical method of closely duplicating what your eye sees.

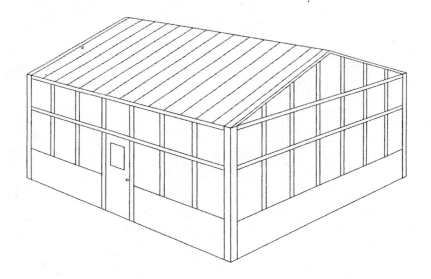

A Two Point Perspective Drawing

In the following illustration, you can see the development of this perspective drawing.

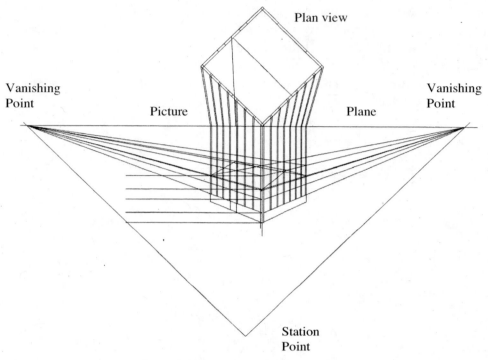

Developing a Two Point Perspective

The plan view was used and set upon a picture plane. Lines were directed from features on the plan view toward the Station Point (where the viewer is standing) to the picture plane. Lines from given heights were directed toward the two vanishing points. The drawing was developed from the combination of these lines.

This is a simplified method of two point perspective. There are other methods that will not be discussed here that allow the drawing to be 'viewed' from different levels and different angles. Perspective drawing is not used for general drafting applications, but reserved for more illustrative presentation purposes.

Oblique drawings

Oblique drawings show the face and two other sides of an object. The face is parallel to the picture plane and the sides are shown with receding lines. Unlike the perspective drawings, the receding lines do not converge to vanishing points. They are drawn at an angle of 30°, 45°, or 60° and like lines are parallel to each other. They may recede up or down, right or left, but the principle is the same.

Unlike the perspective drawing, an oblique drawing is drawn with true measurements on all sides. Like the perspective drawings, oblique drawings give an impression of seeing the object as a normal eye view but because the lines do not converge, it may look somewhat distorted. They do serve the purpose of giving the impression of a 3-D image.

Oblique drawings can be used in conjunction with orthographic views or stand alone. If they are used alone, they are often dimensioned and also may be drawn as sectioned views. Vertical and horizontal lines are easily measured. Since it would be impossible to draw a specific angle in the oblique view, angles are drawn by locating the points of each end and connecting those points with an Oblique line. Circles are drawn as ellipses and can be constructed with four compass arcs or by using ellipse templates.

There are two types of oblique drawings – **Cavalier** and **Cabinet.**

Cavalier Drawings

Cavalier drawings follow the rules described above. They are drawn with a face parallel to the picture plane, use receding lines at the angles of 30°, 45°or 60°, and like lines are parallel to each other. They are fairly simple to draw and are provide an impression of 3-D drawings.

Cabinet Drawings

Cabinet drawings are oblique drawings drawn as described above, but with the receding lines foreshortened to help relieve the distortion. The drawing is made using true dimensions for the face and 1/2 the true dimensions for receding lines.

The following illustrations will show the difference between the types of oblique drawings and how they are developed.

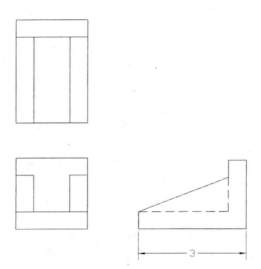

The oblique drawings below were created using the object above. The length dimension was shown to illustrate the resulting 'true measurement' on the oblique drawing at different angles. Although not all the dimensions are shown on the drawing, they were transferred full size as well, and the angled lines were drawn by measuring and locating the distances vertically and horizontally, and then connecting those points.

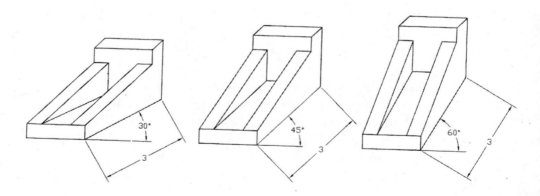

Cavalier drawings

The second type of oblique drawing is the cabinet drawing. The drawing shown below is a cabinet drawing of the same object as in the previous drawings, but the receding dimension has been shortened. The dimension shows that it is one half the actual distance. All other distances are the same, but by shortening all receding lines, there is less distortion in appearance. Note: if it were dimensioned in actual use, it would have the true dimension rather than the half dimension shown here. This dimension was used only to indicate that it was halved

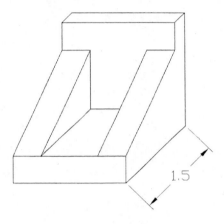

A Cabinet Drawing

Isometric Drawings

Isometric drawings are the most commonly used pictorial drawings. They are used regularly in conjunction with orthographic drawings to give added clarity to the drawing. They give the impression of 3-D but without the complicated construction of perspective drawing. Isometric drawing uses an axis of 120°-120°-120° as shown below. The axis is easy to construct because the right and left arms are 30° from horizontal, easily drawn with your 30/60 triangle. The third leg is vertical. The 30° may be either above or below horizontal, depending on whether you want a view of the top or bottom of the object. The example given here shows the top of the object.

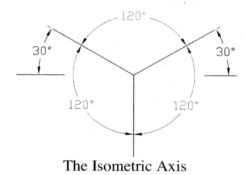

The Isometric Axis

The drawing is started with a boxed sketch – lines parallel to the axis. See the illustration below.

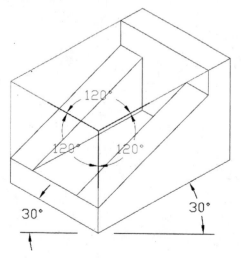

Developing an Isometric Drawing

Using the same object as used earlier in the oblique drawings, you will see the similarity, and the difference of the angles. Note that there are two receding directions – following the axis lines. The development follows the same procedure as in the oblique. Angled lines within the object are located by measuring to points at each end and then connecting these points.

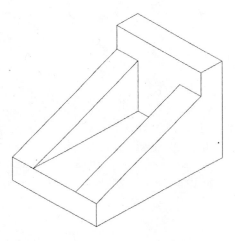

The Finished Isometric Drawing

Isometric drawings do tend to look distorted if the receding lines are too long, but it is the least objectionable distortion for most pictorials. Isometric drawings are drawn using true measurements and may be dimensioned or even sectioned.

Shown to this point, all examples are rectangular, but often there is a need for arcs and circles to show holes or curved edges.

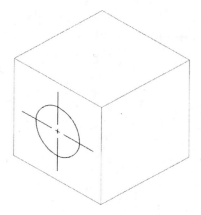

An Isometric Block with a Hole

Since all sides are oblique surfaces, regular circles cannot be drawn. Circles must be drawn as ellipses. The below illustration will show how this can be done.

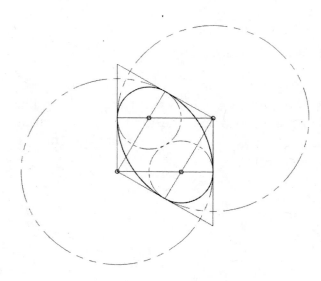

The 'circle' to be drawn is outlined with an isometric square. This square is divided with lines as at left. Lines from the center of one side to a vertex and repeated on the other side provide 'centers' for arcs. The ellipse is made up of four arcs formed by two small compass swings and two large compass swings as shown.

Constructing an Isometric Ellipse

Each side of an isometric drawing must be approached the same way for an ellipse, but each lays in a different plane so must be laid out on that plane as shown below.

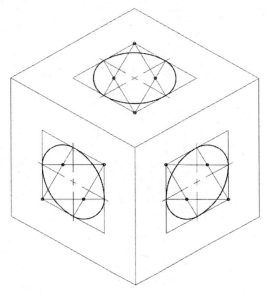

The size and location of the curve or hole dictates the size and location of the square layout, but the procedure is the same.

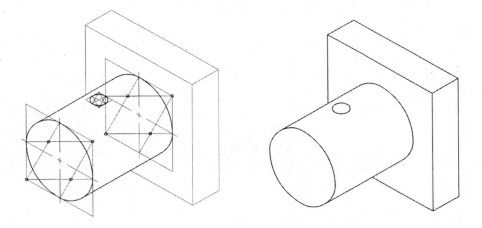

The drawing on the right was drawn using the development shown on the left. Any curved surface could be developed the same way.

SKETCHING

SKETCHING

Now that you are familiar with all of the drawing techniques, understand orthographic projection, and pictorial drawings, you are equipped to draw sketches. Sketching is normally thought of as 'freehand' drawing, which it may very well be, but, also, sketching may be done with the use of templates and straightedges. Sketching overlooks the drawing rules of precision and finesse allowing the drafter to plan or record important information quickly in any setting. Sketching is often thought to be the ultimate in drafting abilities. It has many uses.

Beginning of a new concept or idea. A sketch is usually the beginning of any new concept or idea. To be able to quickly put a thought on paper in a way that others can understand it is an art.

A quick planning tool. The art of sketching is a quick planning tool. A good drafter will often sketch the necessary views as a part of his/her plan before starting a drawing. With the knowledge of what is needed for a good working drawing, a drafter can develop the plan for his/her final drawing in quick sketches before going to the board.

A needed change. A sketch may be the response to a needed change. In the field or in the office, an engineer, designer, architect, or technician will often relate to the drafter with a sketch, when something needs to be altered or redesigned.

In the field. Many situations require that you be able to draw without the board and instruments. A drafter can go into the field and intelligently record information about an object that needs to be drawn. Sketches can be drawn of the necessary views with dimensions and a sketched pictorial can be added for better understanding. This sketch can then be brought into the drafting room and the drawings redrawn mechanically on the board or computer to produce finished drawings.

Document information. Sketching is usually used to quickly capture an idea or to document the shape, size, and appearance of an existing object. Sketching is not done to scale or measured. It is often dimensioned, however, and it is important to record all of the information possible. A good sketch will be drawn in proportion and be as detailed as possible.

Since sketches are often done in the field without the aid of anything more than a pad or a clipboard, it is good to know some techniques for freehand sketching. The illustrations on the following pages will give examples and instructions for sketching techniques.

Sketching Techniques

Sketching uses the same principles and techniques as you have learned in previous chapters. Line conventions are important – centerlines, hidden lines, and object lines will help to convey the thought even though it may not have the same perfection or accuracy. Accurate measurements are replaced with 'proportions' to give a clear image.

Remember, as you sketch, your 'drawing board' is probably a clipboard or pad and so is quite mobile. Turn the 'board' as you work, in a direction that works best for you for each line – hence vertical and horizontal lines can all become temporarily 'horizontal' in the sketching process. Most right-handed people find it easiest to draw from left to right. Left-handed people may want to reverse this. Turn the board for each line if necessary.

Sketched lines are usually more successful if drawn in a series of short strokes rather than attempting one continuous line.

To sketch a straight line, locate the ends of the line first with dots. Start at one dot and aim for the second. The goal of the second dot will help keep the line straight.

Circles and arcs can be sketched by using a square to delineate the size of the circle and dividing the square into sections – first quarters (center lines) and then intermediate lines, etc. place dots or ticks on each line equal distance from the center and then connect the dots with small arcs.

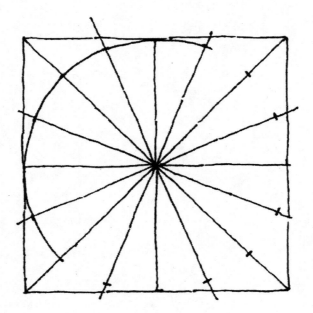

Any curved line can be sketched using the same procedure – by locating the center and radiating rays out from it. Add dots to these rays at the desired distance and connect the dots. It is much easier to visualize these distances with dots than to try to draw the curve without the aid of this beginning.

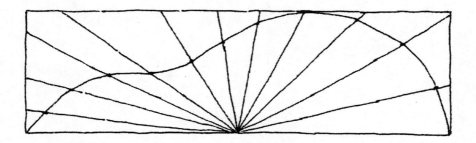

Angles can be 'measured' by 'educated estimation'. Using things you already know, you can create angles. A 45° angle is the diagonal of a square. Lesser angles can quickly be drawn by 'Halves' as below on the left. Half of a 45° is 22-1/2°. Half of 22-1/2° is 11-1/4°. Adjusting up or down on your sketch, they can be used for near angles. 30° and 60° can be achieved by using a diagonal across an arc as shown on the right. Divide the arc into three equal parts and draw to these divisions.

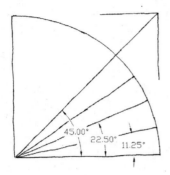
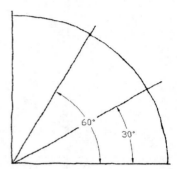

Applying this knowledge, use the same procedure that you have learned in mechanical drawing.

Visualize the end result
Lightly block in the object
Locate center lines
Draw arcs and circles
Draw the rest of the features
Darken the desired lines

Sketching is begun with light construction lines which will give the basic shape and location of features: First, lightly block in the object.

Use dots to locate 'corners' of the block and sketch from dot to dot as in the following drawings.

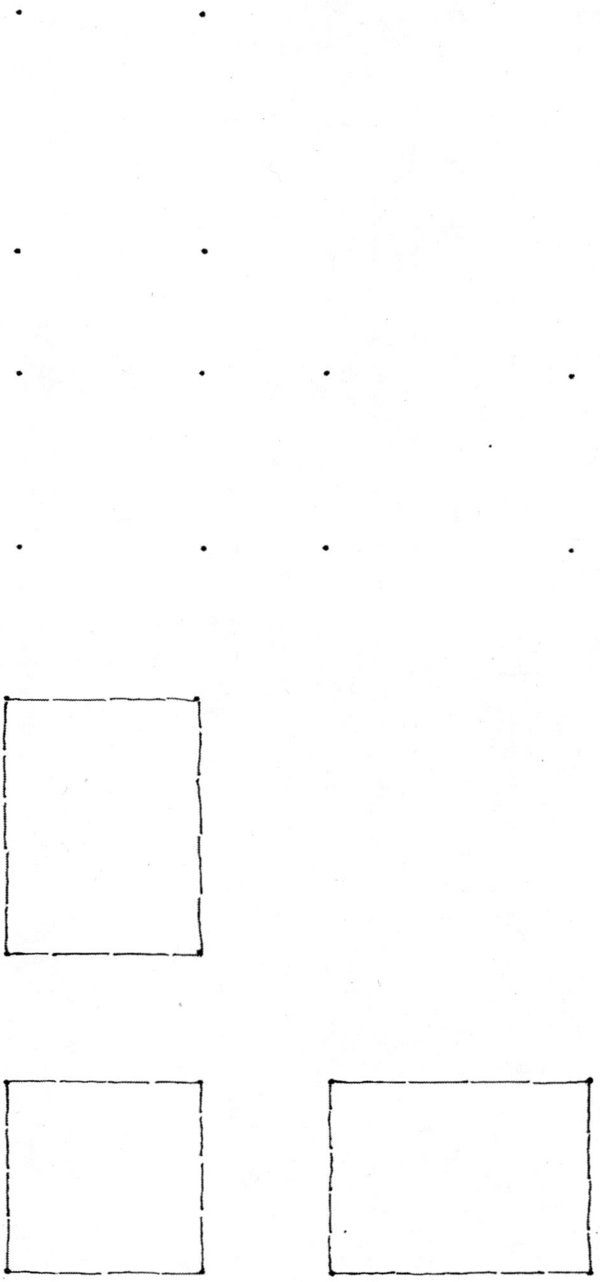

Centerlines are drawn first to locate symmetrical features. On the centerlines, mark off ticks to use as 'goals' for the arcs you will draw. Add intermediate lines between the centerlines and mark them with ticks. Use as many intermediate lines as you need to give enough marks for a curve.

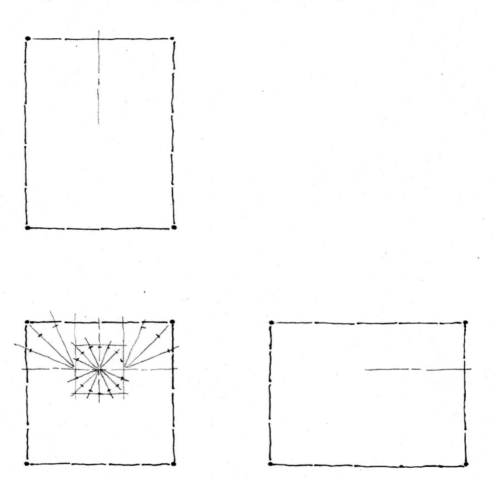

When you have located the points on the arcs and circles, sketch arcs from tick to tick. Then add any other features to the drawing. Hidden lines can be projected from the circle as below. Offsets or other features are added as in a mechanically drawn view, projecting information from one view to another.

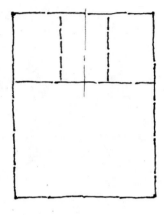

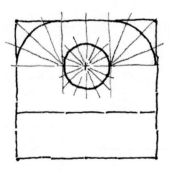

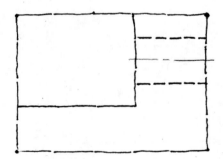

Go back and darken the desired lines as in mechanically drawn drawings. Sketch as many views as necessary to completely describe the object. Erase if necessary. Add dimensions as you would in a mechanically drawn drawing. You can add a sketched pictorial if it will help convey the information.

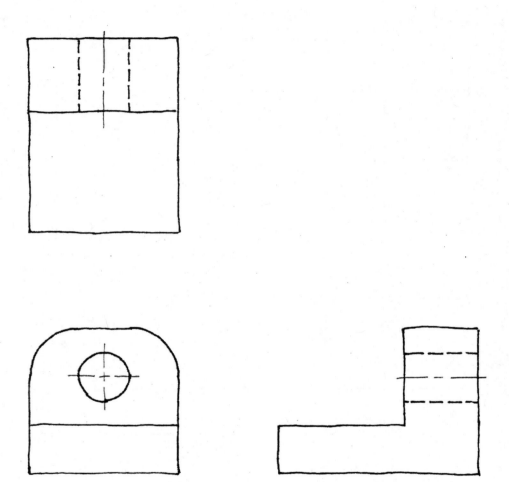

Remember, sketching is no different from mechanical drawing, except for the absence of the instruments. Use the same principles, techniques, and rules to create drawings that are easily read and understood, providing the information that is necessary.

DRAWING EXERCISES

DRAWING EXERCISES

The following pages are exercises to use as assignments / practice. They can be used individually or in a classroom as assignments. Many of the problems could be used in several ways. Copy them as they are or to a different scale. Some require interpretation to develop multi-view drawings from an oblique. Some multi-view drawings could be opportunities to draw isometric or oblique drawings. Most could be opportunities for dimensioning practice.

Use all of the information gained earlier in the book to develop the drawings in an appropriate border on an appropriate sized sheet, to scale or full size. Use proper line conventions and lettering. Practice for quality and accuracy.

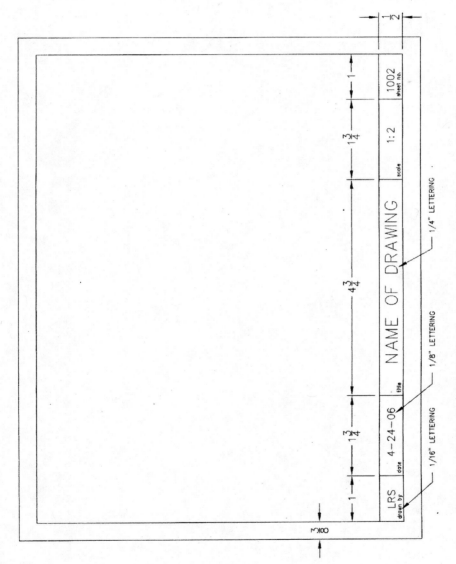

Copy the border and title block on an 8-1/2 x 11 sheet for your drawings using the dimensions given – or create your own using this as an example.

Cam

Trace the dots on a piece of vellum and use a French curve to draw a smooth continuous curved cam outline.

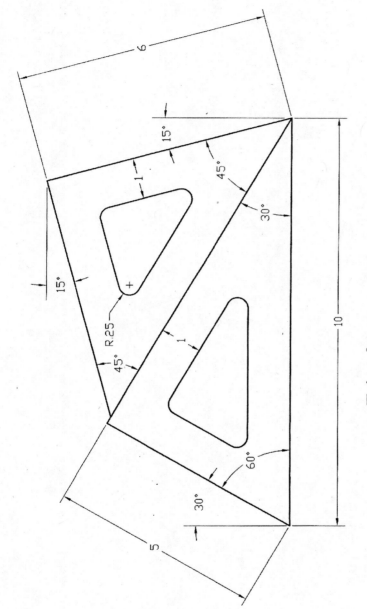

Triangles

Copy the triangles to help you become familiar with the angles that are achievable. Note the outside angles that are shown.

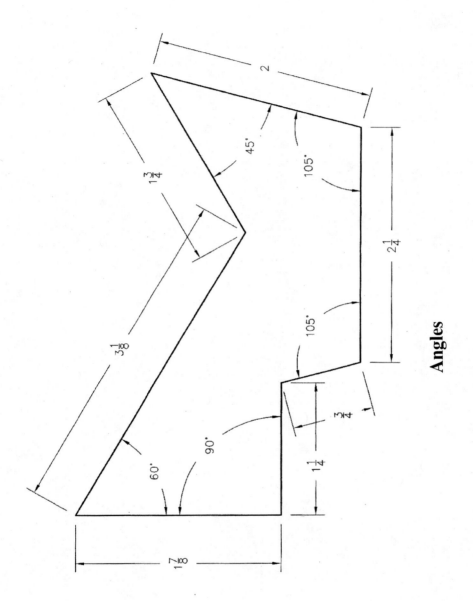

Angles

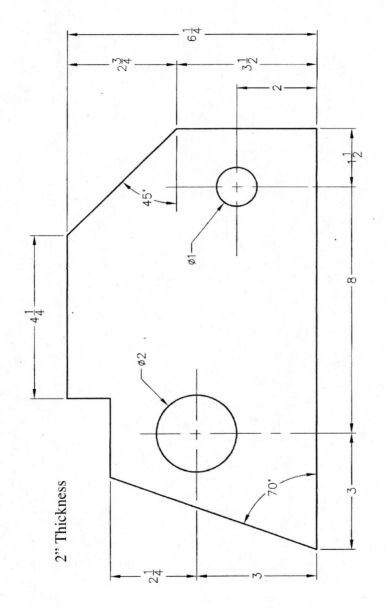

Plate

Can be copied or used for multi-view drawing or pictorial

2" Thickness

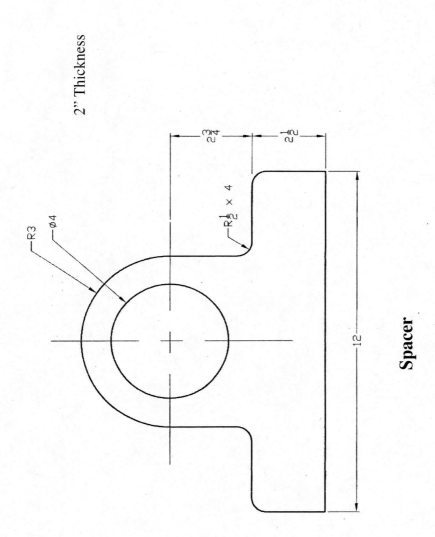

Spacer

113

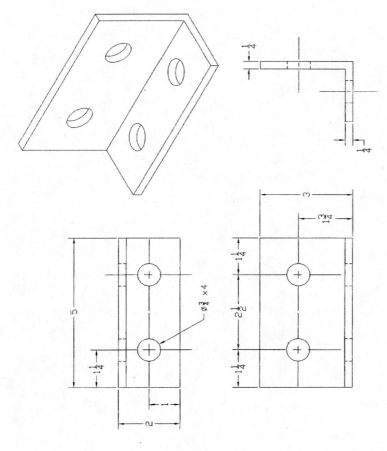

Bracket

115

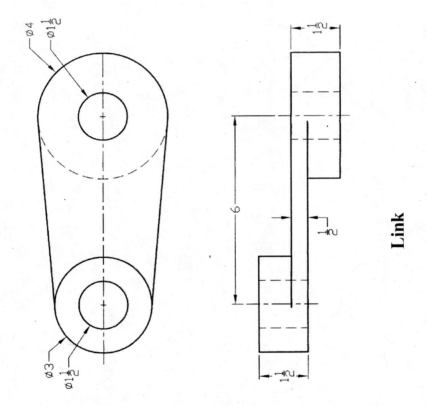

Link

Copy the valve hand wheel and consider a section view.

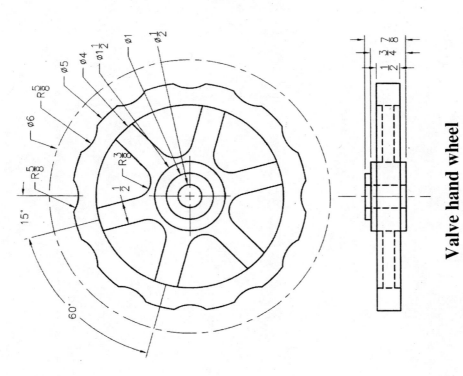

Valve hand wheel

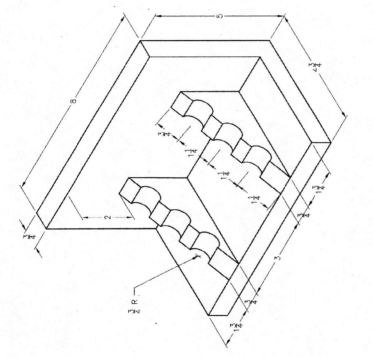

Bookend with pencil rack

Draw working drawings of the bookend with pencil rack.

121

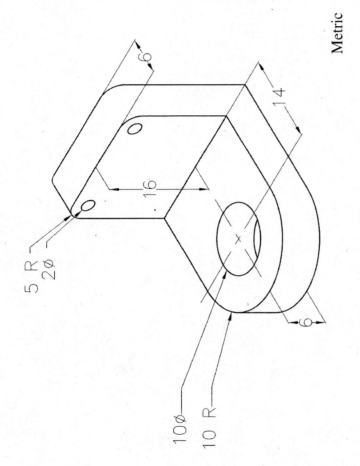

Metric

5 R
2∅

16

14

6

6

10∅

10 R

Hanger

Draw working drawings of
the hanger

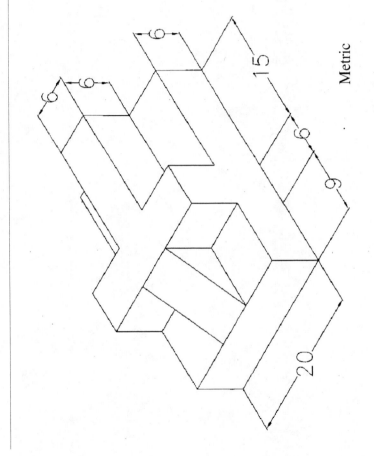

Base Block

Draw in multi-view drawings to a scale that will fit on a chosen sheet of paper.

Metric

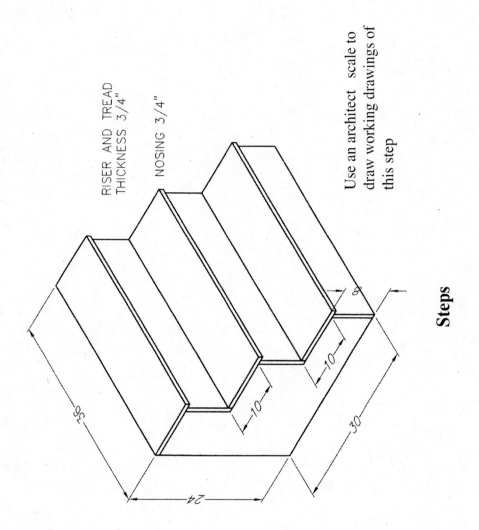

RISER AND TREAD THICKNESS 3/4"

NOSING 3/4"

Use an architect scale to draw working drawings of this step

Steps

Draw the laundry room plan to Scale – 1/2" =1'

The room is 10'x10'.

The door is 36"

The washer and dryer are 30"x30" square.

The laundry tub is 24" square.

The water heater is 18" Dia.

L.T.

W

D

W.H.

24x64 COUNTER

Laundry Room

Draw the office space to
Scale - 1/4" = 1'-0"

Windows – 48" wide

Doors – 36" wide

Dimension the drawing

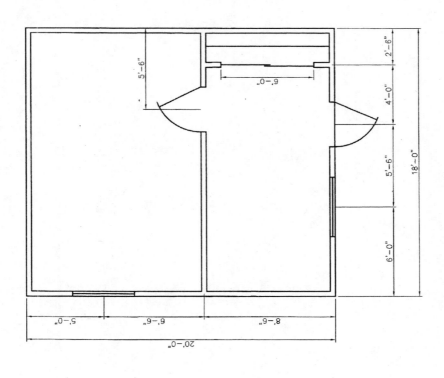

Two-Room Office

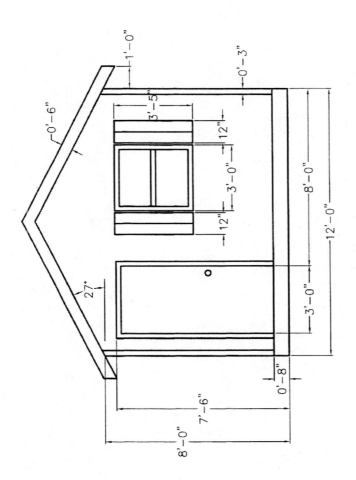

Garden Shed Copy the garden shed elevation using the dimensions given.

Draw to scale – 1/4"=1'-0"

APPENDIX

APPENDIX

The following pages offer some information that may be useful as helps in drawing situations. It is intended to be accurate, but by no means complete. There are many handbooks and more advanced textbooks that will be add much more detailed information for specific and advanced applications.

Decimal Equivalents

Fraction	Decimal	Fraction	Decimal
1/64	.015625	33/64	.515625
1/32	.03125	17/32	.53125
3/64	.04875	35/64	.546875
1/16	.0625	9/16	.5625
5/64	.078125	37/64	.578125
3/32	.09375	19/32	.59375
7/64	.109375	39/64	.609375
1/8	**.125**	**5/8**	**.6250**
9/16	.140625	41/64	.640625
5/32	.15625	21/32	.65625
11/64	.171875	43/64	.671875
3/16	.1875	11/16	.6875
13/64	.203125	45/64	.703125
7/32	.21875	23/32	.71875
15/64	.234375	47/64	.734375
1/4	**.250**	**3/4**	**.750**
17/64	.265625	49/64	.765625
9/32	.28125	25/32	.78125
19/64	.296875	51/64	.796875
5/16	.3125	13/16	.8125
21/64	.328125	53/64	.828125
11/32	.34375	27/32	.84375
23/64	.359375	55/64	.859375
3/8	**.3750**	**7/8**	**.8750**
25/64	.390625	57/64	.890625
13/32	.40625	29/32	.90625
27/64	.21875	59/64	.921875
7/16	.4375	15/16	.9375
29/64	.453125	61/64	.953125
15/32	.46875	31/32	.96875
31/64	.484375	63/64	.984375
1/2	**.5000**	**1**	**1.000**

Basic Abbreviations used on Drawings

ASSY	Assembly	MED	Medium
AUX	Auxiliary	MID	Middle
AVG	Average	MIN	Minimum
CLG	Ceiling	MISC	Misc
C to C	Center to center	NTS	Not to Scale
CL	Centerline	NO.	Number
CIR	Circumference	OZ	Ounce
CONC	Concrete	OD	Outside Diameter
CSK	Countersink	PKG	Package
Cbore	Counter bore	POS	Positive
CYL	Cylinder	PSI	Pounds per Sq.In.
DET	Detail	IN	Inch
DIA	Diameter	QUAL	Quality
DIM	Dimension	QTY	Quantity
DWG	Drawing	R or RAD	Radius
EA	Each	REV	Revision
ELEV	Elevation	RPM	Revolutions per Minute
EST	Estimate	MAX	Maximum
FIG	Figure	MIN	Minimum
FIN	Finish	SCH	Schedule
FTG	Footing	SCR	Screw
GA	Gage	SECT	Section
HEX	Hexagon	SH	Sheet
IN	Inch	SF	Spotface
KWY	Keyway	SQ	Square
LAB	Laboratory	STD	Standard
LH	Left Hand	TAN	Tangent
LG	Length	THK	Thick
LTR	Letter	TYP	Typical
LIN	Linear	VAC	Vacuum
LG	Long	VAR	Variable
MAJ	Major	VERT	Vertical
MFG	Manufacture	WP	Weather Proof
MAX	Maximum	WT	Weight
MECH	Mechanical		

Measurement conversions

WHEN YOU HAVE	MULTIPLY BY	TO FIND
Inches	25.4	Millimeters
Inches	2.54	Centimeters
Feet	0.3048	Meters
Feet	30.48	Centimeters
Yards	0.9144	Meters
Miles	1.609	Kilometers
Square inches	6.4516	Square centimeters
Square feet	0.0929	Square meters
Square yards	0.8361	Square meters
Square miles	2.60	Square kilometers

Millimeters	.04	Inches
Centimeters	.4	Inches
Meters	3.3	Feet
Meters	1.1	Yards
Kilometers	.6	Miles
Square centimeters	.16	Square inches
Square centimeters	.001	Square feet
Square meters	10.8	Square feet
Square meters	1.2	Square yards
Square kilometers	.4	Square miles

Formulas

Note - Pi = 3.1416

Circle:	$D = 2R$ Area = Pi R^2 Circumference = Pi D
Rectangle	Area = L x W
Parallelogram	Area = bh
Trapezoid	Area = $H/2$ (b_1 + b_2)
Triangle:	Area = ½ Base x Height
Ellipse:	Area = Pi long radius x short radius
Cube:	Volume = Area B x Height
Cylinder:	Volume = Area/base x Height
Pyramid	Volume = (1/3) Base x Height
Cone	Volume = (1/3)area of base x Height

Basic Materials Symbols used in sectioning

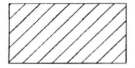

Cast Iron/ generic symbol

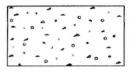

Concrete

Steel

Wood, grain

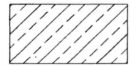

Brass, Bronze, Copper

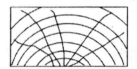

Wood, end grain

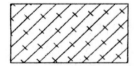

Aluminum / Magnesium

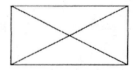

Rough wood end grain

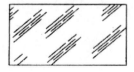

Glass

Earth

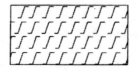

Leather/ Fabric

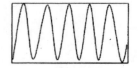

Insulation

Basic Architectural Symbols

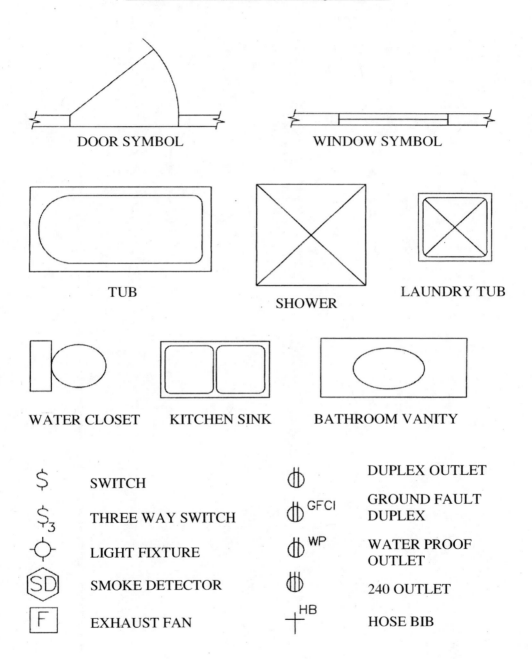

DOOR SYMBOL

WINDOW SYMBOL

TUB

SHOWER

LAUNDRY TUB

WATER CLOSET

KITCHEN SINK

BATHROOM VANITY

$ SWITCH

$3   THREE WAY SWITCH

LIGHT FIXTURE

SD   SMOKE DETECTOR

F   EXHAUST FAN

DUPLEX OUTLET

GFCI   GROUND FAULT DUPLEX

WP   WATER PROOF OUTLET

240 OUTLET

HB   HOSE BIB

Sheet sizes

Standard sheet sizes

	Inches	mm
A	8-1/2 x 11	215 x 279.4
B	11 x 17	279.4 x 431.8
C	17 x 22	431.8 x 558.8
D	22 x 34	558.8 x 863.6
E	34 x 44	863.6 x 1117.6

ISBN 141209676-6

9 781412 096768